DREADNOUGHT

Britain's First Nuclear Powered Submarine

By Patrick Boniface

With a Foreword by Rear Admiral Paddy O'Riordan

Periscope Publishing

Penzance

First published in 2003 by

Periscope Publishing Ltd.

33 Barwis Terrace

Penzance

Cornwall TR18 2AW

A CIP record for this book is available from the British Library

ISBN No 1-904381-09-X

Printed in England by Antony Rowe Ltd

Eastbourne

Table of Contents

Illustrations

The publishers would like to thank the following for their permissions to use the photos in this book. Our best efforts have been made to trace copyright holders. We have indicated where this hasn't been possible.

Royal Navy Submarine Museum, Gosport, pgs 41, 42, 44, 45, 46, 47, 50 and front and rear covers

Navpic, pgs 46, 48

HMS *Maidstone*, pg 43

The Crown, pg 49

HMS *Neptune*, pg 48

Foreword

What's in a name?

This is the story, in fact more of a log, of the life of the Royal Navy's first nuclear powered submarine HMS *Dreadnought,* the ninth ship to bear this famous name, the first having been built in 1573.

The book provides a most fascinating insight into the aspirations, frustrations, failures and successes surrounding the advent of nuclear power in the Royal Navy, the biggest and most significant change since the transition from sail to steam.

It is all about pioneering, planning and achievement in the face of enormous challenges. It is a testimony and record of human endeavour in making what at times seemed the impossible become a reality. One can imagine the excitement of being the first British submarine to be 'underway on nuclear power', of being the first British submarine at the North Pole and a record-breaking 43,000 mile 194-day deployment.

Dreadnought was all about people from so many professions and trades coming together to make it all happen...*Dreadnought* was a people submarine with a great sense of history and tradition backed up by a fine motto; "Fear God and Dread Nought".

Dreadnought and all those who worked for her, on her or served in her did our nation and the Royal Navy proud.

Rear Admiral Paddy O'Riordan.

Acknowledgements

In the course of writing this history of HMS *Dreadnought* I have been indebted to a great many individuals and organisations for their expertise and for their kind loan of materials and information. Amongst those who require special mention are: Dave Tull, Chatham Historic Dockyard, everyone at the Submarine Museum at Gosport, Mike Smith at BAE Systems at Barrow, all the staff at Portsmouth City Council Library Services, and the staff at the Public Records Office, Kew.

Patrick Boniface,

January 2003

Chapter One

Design and Construction

The Nuclear age saw a great deal of military aspirations; one of the most important was the prospect of powering naval vessels by nuclear power. The ideal candidates for such a power source were submarines where extremely long endurance would be an immensely useful attribute. A nuclear power plant would afford any submarine so equipped almost unrivalled flexibility to remain underwater for long periods of time and to attack almost at will. The Americans were the first to harness the splitting of the atom for this purpose when the world's first nuclear powered submarine USS *Nautilus* commissioned. The Royal Navy, however, also viewed the potential as well worth investing time, energy and money into.

The Royal Navy had successfully come through World War Two but many of its warships were of old and obsolete design, especially when compared with the latest captured examples of German U-boats. Added to this the desire to experiment with new designs and technologies was too much to resist. Accordingly designs were first mooted in the late 1940's for a submarine that could incorporate the use of gas-cooled nuclear reactors, similar to the research reactor at Calder Hall. This particular reactor was, however, overly complicated and cumbersome in design and would be impracticable onboard a submarine that was subject to the movements of the sea. Initial designs for a British nuclear submarine in the early 1950's stalled through the immaturity of the available technology.

If nuclear fuel was out of the frame for a while, then alternative fuels were certainly considered. The years immediately after the Second World War saw the Royal Navy experimenting with what was seen as the fuel of the future. This was the highly explosive and corrosive substance hydrogen peroxide. The captured German submarine *Meteorite* used this fuel and in addition two British designed boats were built, *Explorer* and *Excalibur* as test beds. These evaluations were conducted to assess the suitability of using hydrogen peroxide onboard submarines, where the fuel was also tested as a power source for future torpedoes. In the end the weaknesses far outweighed the merits as anyone who served onboard the two 'EX' class submarines in the 1950's will testify. Also the torpedoes using this fuel were quickly found to be lethal and not just to the enemy. Indeed it is widely suspected that torpedoes of this type may have caused the loss of the Russian submarine *Kursk* in 2000. HTP torpedoes also accounted for the loss of *Sidon* alongside her depot ship *Maidstone* in the 1950's.

The complexities of the engineering necessary for marine nuclear propulsion required huge investment in terms of time, money and manpower. Lord Carrington recalls that, "As First Lord of the Admiralty I was involved in making all the preparations and negotiations necessary to get it going. The most important thing was, of course, the co-operation of the Americans because we didn't really know how they managed their nuclear reactors." He also remembers that, "It was an extremely complicated business to build a submarine of that size. I think the technical problems with doing it were greater than we had anticipated but they were all overcome." Accordingly leading British companies were brought in to work together as a consortium on the submarine project. Amongst many companies the principal ones included Vickers Armstrong's, Engineering (VAE), Foster Wheeler and later on in the programme Rolls Royce also joined the consortium. No one company had all the necessary experience or expertise to complete the submarine project on its own and each brought to the table specialists in fields such as shipbuilding, engineering, reactor chambers and steam plants. The project was initially supervised by Captain Harrison Smith who was appointed in 1955 to head the Naval Section of the Atomic Energy Authority at Harwell in Oxfordshire. Captain Smith had at his disposal both military and civilian scientists, engineers and specialists to work on the nuclear submarine programme.

One of the first decisions made was to evaluate British nuclear technology and to see how easy or indeed practicable its installation and maintenance would be onboard a mobile submarine. For this reason a shore based nuclear reactor facility was to be built to design, test and evaluate the British nuclear power plant.

The government of Anthony Eden in 1956 gave final go-ahead for the construction of this test reactor at Dounreay in the extreme north of Scotland. The site chosen is one of the most remote parts of the United Kingdom. Much of the reasoning behind this was for security; other considerations were more practical in nature, such as the ease of access to large amounts of water to cool the reactors. Work soon started on the Dounreay Submarine Prototype (DS/MP) reactor. 1956 also saw the Eden government initiate detailed studies into using emerging technologies for possible installation into a British designed and built submarine. Specialist personnel were assigned to the project including Rear Admiral (E) Guy Wilson who became Rear Admiral Nuclear Propulsion. Admiral Wilson had joined the Royal Navy as a cadet in 1920 and studied at the Naval College in Dartmouth. He was promoted to the rank of Commander in 1940 and from 1941 to 1943 served onboard the cruiser Berwick as Engineer Officer. In 1943 he joined the Engineer in Chief's department and assisted in designing the machinery spaces in the aircraft carriers *Ark Royal* and *Eagle*. After the war in 1946 he spent three years as First Assistant to the Manager of the Engineering Department in Portsmouth Dockyard. 1949 saw promotion to Captain and he assumed the role of deputy director of the Department of Aircraft Maintenance and Repair which moved to RNAY Fleetlands, Gosport. On 12 December he was promoted to the rank of Rear Admiral and became Deputy

Engineer in Chief for Fleet Maintenance and Administration. He had been brought onto the *Dreadnought* programme to steer it as the programme was seen to be stalling somewhat, but Admiral Wilson soon got the project back on the right track.

It is fair to say that British enthusiasm for the nuclear submarine project exceeded British experience of designing nuclear reactors and their associated sub systems. This led to a great many design alterations on numerous occasions that caused frustration. All the while, across the North Atlantic, the US Navy quickly expanded its fleet of nuclear submarines and also further developed the capabilities of their pressurised water reactors. At the time British engineers were denied access to what was, (and still is), highly classified information about US submarine reactor design. They struggled to make the British system fit the confines of a submarine's hull and all this time the schedule for Britain's first nuclear submarine began to slip. In some quarters the British decision to go it alone was being questioned. Was it time to go to the Americans and ask for their assistance? The official view started to become that British pride might have to take a beating but the programme had to be successful.

Across the North Atlantic the US Navy's submarine force was headed by the formidable presence of Admiral Hyman Rickover. This charismatic man had through perseverance forced nuclear power into the USS *Nautilus* and had changed forever the shape of submarine warfare. Rickover was a hard man who was willing to make enemies in order to achieve his goals for the US Navy. What he wanted was for the US Navy to be the world's premier navy and through nuclear power he saw a means to achieve this. Originally Rickover was reluctant to help the British with the project but was persuaded by a number of meetings with Lord Louis Mountbatten, the First Sea Lord. Subsequently in 1956 Admiral Rickover invited the British team to see what he had achieved.

In 1956 the British design team investigated a number of submarine designs with different reactor configurations. One of the first design characteristics set was that the hull shape would be along the lines of the USS *Albacore*, a streamlined teardrop or cigar shape that looked something like a huge whale. The British team had also through evaluating the merits of various layouts come to the conclusion that the best layout would closely resemble that used in the USS *Skipjack*. Indeed the Americans were so alarmed at the similarity in design concepts to the *Skipjack* that they believed that espionage could have been used to obtain such classified information from the United States. Of course, none had been, it was simply the most logical and best location for the system to be located within a submarine's hull.

The end of 1956 saw The First Sea Lord (Lord Mountbatten) recommending Sir Victor Shepheard to head the Nuclear Submarine Programme. It was also around this time that a breakthrough was first brokered between Britain and the United States. Admiral Rickover softened his viewpoint about exporting US nuclear technology and suggested to Lord Mountbatten that he would sell a reactor plant to Britain in order to get the programme fully back on track. In February 1958, Rickover visited Britain and pushed hard to get Rolls Royce a £10 million contract with Westinghouse for the delivery of a complete S5W reactor and associated machinery, including guidance on the setting up of a plant to make S5W cores. There was, of course, a raft of conditions to the deal, principal amongst them that the reactor remained American property and that its design and operation remain a closely guarded secret. The decision also had to be ratified by the American government and required an amendment to both the US constitution and to the American Atomic Energy Act. Britain, faced with technological problems, which whilst surmountable given sufficient time, were going to surely delay the construction of the submarine, the MacMillan government signed this bilateral UK/USA agreement. This came to be known as 'The 1958 Agreement' and remains current to this date. Thus once the agreement was signed an S5W reactor was procured for the British nuclear submarine programme and this kept the programme on time and budget.

Following its acquisition the S5W reactor presented some problems for the engineers to solve. Its size and shape were optimised for the American *Skipjack* class not for the British design which was quite advanced at the time. Indeed, it was reported in official communications that, 'Except for the fact that the Atomic Energy Authority's operation in this country involves the use of various materials under irradiated conditions, the type of reactor fitted in the submarine is so different from any AEA reactor that the experience gained by the AEA in the engineering field has not been of great assistance to the Admiralty because it has not been applicable." This not withstanding, the S5W reactor was going to be fitted in *Dreadnought* and it was essential that the structure of the submarine be strong enough to accommodate it safely. Professor Douglas Faulkner remembers working on the design of *Dreadnought*. "Because the US Navy's *Skipjack* class of nuclear attack submarines were structurally poor the RCNC decided that for *Dreadnought* we should pay more attention to fatigue design at axi-symmetric stress concentrations such as the junctions of conical and cylindrical portions of the pressure hull brought about to cater for the nuclear plant."

Engineers pored over their drawing boards in an effort to marry the two halves of the hull together. The forward section, including the sonar rooms, torpedo rooms, control sections, fin and accommodation were to be of purely British design. This British section was designed by L J Rydill who took great care to design it to accommodate the latest Type 2001 sonar. Rydill also brought into the hull new standards of accommodation in addition to new equipment for ballasting and diving control, ventilation and for controlling the atmosphere inside the submarine. *Dreadnought* would also introduce the water ram torpedo discharge system. The aft section, however, was of mostly American design, although it would be constructed in Britain. In appearance the submarine design showed a pronounced draw down on the spine of the submarine where the two sections met. The after end of *Dreadnought* after Number 44 bulkhead became known as the 'American Sector' whilst the bulkhead between the two was christened 'Check Point Charlie', named after the gateway between the US sector and East Berlin during the Cold War.

As designed this American Sector was divided into four main compartments; the Reactor Compartment (RC), Tunnel, Auxiliary Machinery Space and Engine Room.

The Reactor Compartment contains the Reactor, the Steam Generators (SG's) or Boilers and other associated equipment. The Reactor is used as the Heat Source for the generation of Steam. This heat is removed from the Reactor by pure water under pressure, and pumped via the Primary Loop into the Steam Generators. Here the Primary Water gives up its heat to the Secondary Water through a heat exchanger. The Primary Water then returns to the Reactor to complete the closed cycle.

A shielded tunnel is provided above the RC to allow access from the forward spaces to the Machinery Spaces aft.

The Auxiliary Machinery Space has two levels. The Upper level contains the Main Electrical Distribution Switchboards and the Reactor Instrumentation, Control and Protection Cabinets which are sometimes known as 'Idiots Alley'.

One other notable item is 'Hammersley Folly' for those crewmembers in need of toilet facilities. Designed by Commander 'Spam' Hammersley, *Dreadnought*'s first Engineer Officer, this item was fitted to provide some 'relief' for the after watch keepers!

The Engine Room contains the Main Motor Generator Sets, Diesel Generator, Salt and Fresh Water Pumps for Reactor Services, and the Main Feed Pumps for pumping the Secondary Water into the Boilers.

The Engine Room is also split into two levels. The upper level contains the Main Turbines, which are coupled to one single shaft via a gearbox. It also contains the Turbo-Generators which produce the electrical power required onboard. *Dreadnought* was also equipped with a sea water distiller which produced the pure water required for the Boiler Feed System. It also supplied potable water for the laundry, shower baths and washing facilities.

In the Lower Level were the Main Condensers, Main Sea Water Pumps, Air Conditioning Sets and other auxiliaries.

The hub of the Machinery Space is the Manoeuvring Room. Here the Reactor, Electrical and Propulsion Plants are controlled. The Manoeuvring Room was always manned when at sea, by at least four watch keepers at all times.

Another key element of the *Dreadnought* design was the installation of a brand new sonar for the hunter killer submarine, the Type 2001 active/passive set. Design of this set was started in 1957 to meet a Staff Requirement for a long range, low to medium frequency conformal bow mounted sonar and was specially designed for use on nuclear submarines by the team at AUWE Portland. Originally the constructors had wanted to fit the Type 177 sonar set to the *Dreadnought* but AUWE considered it possible to design and develop a new Sonar system in time for *Dreadnought*'s commissioning in 1962. However, it was necessary for the constructors to accept a change in the design of the submarine to accommodate a bulge in the bow caused by the sonar's transducer array. Normally a project of this scale takes between seven and ten years from conception to delivery so a streamlined schedule was developed. Despite the pressure of time the design team at AUWE was quite small just fifteen people in total plus those from industry and other departments. Cottage Industries (a hybrid Organisation paid for by the Admiralty and run by Plessey Company) was used as an AUWE extension and much detailed design work was carried out there. The engineers responsible for the Type 2001 sonar decided that to design the set within the five year timetable certain short cuts had to be taken. The first was to break the system up into its sub systems. Type 2001 sonar consisted of several sonar sets using a single transducer array. An engineer was also appointed to develop and monitor the manufacture of that part. This unfortunately led to a division of 'know how' among the Technical experts and only two of the project engineers had detailed knowledge of the whole set, S Mason and J Marchent. On projects like this there are usually five stages to be met, a) Breadboard model, b) Experimental model, c) Prototype, d) Pre production model and e) Production model. Each of these phases is built and tested either in the Laboratory or at sea and the lessons learned are incorporated into the next model. When this project was started it was realised that there was no time for lengthy research programmes into new techniques and most of the design was based on known and proven techniques. This lessened the calculated risk that was taken when the full development schedule was telescoped. There was no breadboard model. New design circuits were mocked up in the laboratory and developed until they gave optimum performance but the first set consisted of an experimental model which was fitted in the sonar trials ship *Verulam*, (a Royal Navy frigate with armament removed), with both naval and civilian personnel onboard. A series of sea trials were held in various parts of the Atlantic and the North Sea. Extrapolating the results of these trials inevitably included a lot of guesswork because the frigate was reluctant to steam at a depth of 400 feet to get below the surface layer.

In spite of these drawbacks the value of the work done onboard *Verulam* was enormous and much time, effort and money was saved. It was not possible to build and test a prototype model. Instead the actual installation in *Dreadnought* has been described as an experimental prototype. This was necessary because a) the manufacture of the submarine set went on concurrently with the frigate trials, and many decisions had to be taken before the answer was forthcoming from the trials and b) the submarine set was the first one to be installed in the limited space and to be operated in the ocean.

Other items that were to be used in the construction of *Dreadnought* were tested onboard the *Porpoise* class submarine *Finwhale* in 1958. *Finwhale* had been sent to the ice cap towards the North Pole. Her diesel electric propulsion wouldn't permit a passage to the Pole itself, but the conventionally powered submarine ventured under the ice for some trials of equipment. These included SINS, (Ship's Inertial Navigational System), brand new periscope heaters, sleeping bags and also a specifically reinforced and strengthened fin. *Finwhale* was escorted on these trials by the 'A' class submarine *Amphion*. *Amphion* was declared as the safety boat and exploded charges near to the ice edge in order for the crew of *Finwhale* to safely and accurately judge their location.

Dreadnought was also equipped with two other sonar types, Type 2007 and 2019. The 2007 unit is a low frequency passive sonar with 24 separate hydrophones housed within the array. These listen for any noises and anything that is detected would automatically be processed by the boats computer to indicate friend or foe. The Type 2019 is similar but can be operated in either passive mode or active mode. It is also known as PARIS for Passive/Active Ranging Intercept System. It provided the same information to the crew of the *Dreadnought* as the Type 2007 but with a higher frequency range.

By early 1959 a great deal of progress had been made in the design and forward planning of the submarine's construction. In January 1959 Admiral Sir William Davis, Commander in Chief Home Fleet stated that, "I attach tremendous importance to the submarine as an anti submarine weapon. I don't think any of us, even on the naval side, realise the revolution in thinking and naval strategy that the advent of the nuclear submarine has made."

The summer of 1959 saw a great deal of speculation about possible names for the submarine - many had been suggested but it took the Earl of Selkirk, First Lord of the Admiralty in July 1959 to announce the name of the new submarine. "We are calling this ship *Dreadnought* because it is opening a new epoch just as was the old *Dreadnought,* built fifty years ago."

By now the US Navy were becoming happier with the prospect of Britain having nuclear submarines and decided to allow a select number of British submariners and technical Senior Ratings to train onboard their submarines in the operation of the S5W reactors and in the handling of Albacore type hulled boats. Accordingly in 1959 the Commanding Officer (designate) of *Dreadnought* Commander Barnaby Frederick Palmer Samborne (known as Peter) and several of his officers and ratings were given six months sea training in US nuclear submarines including USS *Skipjack,* USS *Skate* and the original nuclear boat USS *Nautilus.* Commander Samborne was born in Somerset and entered the Royal Navy at Dartmouth in 1938 at the age of 13 and a half. During the war he was serving onboard the cruiser Liverpool when she was torpedoed and Samborne was mentioned in dispatches during this incident. He would later serve onboard the battleship *Malaya* and the *Laverton* before starting a career in submarines from early in 1944 when he joined the *Amphion* at Barrow in Furness. He would later serve in HM submarines *Alliance, Sturdy* and in command of *Tabard* and *Trenchant* at Malta. From January to March 1960 he attended the Royal Naval College at Greenwich for the 1st Nuclear Course along with Lieutenant Commander Squires before further training in the States onboard the USS *Skipjack.* After time spent at Flag Officer Submarines HQ and at the Royal Naval College in Greenwich he took up the post of commanding officer HM submarine *Dreadnought.*

By 1958 Vickers Armstrong's Ltd at Barrow in Furness had been selected as the chosen shipyard to build *Dreadnought,* having probably the best record of submarine construction in the United Kingdom. Press and TV crews jostled with crowds of shipyard workers to witness the historic first moments of the submarine.

HRH Prince Philip arrived at the shipyard on 12 June 1959 to perform the official laying down of the first steel and was greeted by an extremely well turned out guard of honour which was duly inspected. The Prince was escorted on a tour of the shipyard that took in a look at the Chilean destroyer *Almirante Williams* which was packed with shipyard employees who had some of the best vantage points to witness the event. He also chatted with Chilean dignitaries who were witnessing the construction of the destroyer. Addressing the crowd of shipyard workers and dignitaries the Duke of Edinburgh said, "There is perhaps the doubtful prospect for future submariners of remaining submerged for the whole period of a two year commission. On Monday, the Queen presented her Colour to the Submarine Command. Today that Colour is here at Barrow with a guard of submariners to witness an event which will most certainly revolutionise the Navy as a whole and the submarine service in particular. It should be abundantly clear by now that nuclear submarines will ultimately transform the sea power of the Atlantic Alliance." Later in his visit HRH Prince Philip moved a handle which caused a gamma beam to operate a winch. This then drew a 30 foot cylindrical section into position on the slipway. This simple action saw the start of work on *Dreadnought*. After this the Duke of Edinburgh fired two cartridge hammers that bolted a commemorative plate to the hull to mark the occasion.

Vickers had built hundreds of conventionally powered and configured submarines up to this point and *Dreadnought* presented something of a challenge to the shipyard and its workers. Vickers, however, rose to the challenge. They did at times experience some problems with new and innovative practices demanded by nuclear safety concerns. One such involved demanding welding techniques. These were found to be particularly difficult around primary and secondary systems. Other problems were experienced with welding quenched and tempered steel in very tight and awkward locations and brazing metal and tubes in sizes not previously attempted. Further frustrations were expressed with machining and bonding lead to steel and other metals. All the while the shipyard had to enforce incredibly stringent cleanliness levels.

Professor Douglas Faulkner recalls visiting the shipyard on numerous occasions to check progress on her construction. "I was visiting Vickers at Barrow in Furness frequently to oversee the installation of about 500 strain gauges on the pressure hull structure and to check the as-built circularity of the pressure hull frames."

"During this period there was a 'scare' that, through a possible paint coding error on the plating, mild steel plating could have been worked into the pressure hull instead of the high alloy heat treated QT35 steel specified for construction. As the difference in yield stress was about 2.3 to 1.0 which was larger than the safety factor, this was a very serious concern and the problem was how could this be checked non destructively in an already completed pressure-hull."

Professor Faulkner continued, "I devised a simple practical test for identifying any mild steel plating; it became known as the 'Faulkner Test'. Its basis was that measurements of Brinell hardness of the two materials consistently showed that QT35 steel was more than twice as hard as for mild steel. My test, therefore, was to sandwich a hardened ball bearing in situ between the external pressure hull and a piece of known QT35 plating, and this piece was struck heavily by a 4 pound hammer. The diameter of the two indentations were measured and would be similar for QT35 pressure hull plating but some 50% bigger on the pressure hull if it was made of mild steel. All the pressure hull plates were screened, and fortunately found to be of QT35 steel."

All throughout the construction of the new submarine the threat of secrets being leaked to the Soviets was taken very seriously, so much so that whenever there appeared to be leaks they were seriously investigated. Usually, they were just rumours and amounted to nothing, but towards the end of 1959 there was one persistent rumour that would not go away. The Admiralty

Underwater Weapons Research Establishment based at Portland was the scene when two employees came under suspicion. The two were Harry Houghton and Ethel Gee who with another three people, Gordon Lonsdale and a married couple called Kroger, had tried to obtain sensitive details about the design and construction of *Dreadnought*. The tip-off about the five came from a Polish intelligence officer who would later defect to the west. The Pole gave details about Houghton, Gee and Lonsdale and when police arrived at the Old Vic Theatre in London on Saturday 7 January 1961 they arrested Houghton, Gee and Lonsdale with a shopping bag full of details about the nuclear submarine. Gordon Lonsdale received 25 years for his part in the conspiracy, whilst Ethel Gee got 15. It was widely accepted that the Soviets had received a great deal of information on *Dreadnought*'s hi tech sonar.

One incident occurred during the later stages of *Dreadnought*'s construction when the fin, which had been fabricated elsewhere in the shipyard was brought to the pressure hull to be bonded with it. An engineer discovered to his shock that the fin was four inches too short. The reason for this was a problem with communication between Admiralty Surface Weapons Establishment and Vickers about the height of the masts. The solution was the insertion of a gusset in the fin.

The *Dreadnought* was light years ahead of existing submarine technology and as designed introduced a great deal of new ideas in submarine operation. *Dreadnought* was designed to be controlled by means of hydraulics utilising a joystick control column. A great deal of technology was taken from the aerospace industry and reconfigured for use in the submarine. One example was the extensive cockpit style instrument panel inside the control room of the submarine. The teardrop hull shape permitted *Dreadnought* to manoeuvre just like an underwater aircraft and also had the equivalent of an 'automatic pilot'.

Submariners are generally a hardy breed who during the war grew accustomed to sometimes appalling conditions onboard submarines. Many of these conditions persisted in the diesel electric submarines of the post-war period, but onboard *Dreadnought* living standards rose spectacularly. Accommodation onboard was of an unprecedented standard. *Dreadnought*'s crew were given undreamt of facilities even in surface vessels. They enjoyed the ability to have a shower, they could also launder their own clothes in a special laundry. There were separate mess spaces for senior and junior rates. These were arranged on either side of a large galley set out as a cafeteria. As *Dreadnought* was expected to stay away for months on end on patrol and underwater for most of the time the physical and mental health of her crew was addressed. Accordingly appropriate decorations and furnishings onboard were fashioned to be both attractive and functional. Furthermore to ease boredom onboard on long voyages *Dreadnought* had cinema equipment, an extensive library and a large number of tape recordings.

The construction of *Dreadnought* proceeded well and it was during the latter stages that a test was devised to ensure that the reactor compartment and associated areas were truly watertight. It was essential that any leaks would to be contained within the confines of the boat.

A large quantity of vitamin C, which smells of oranges, was pumped into the machinery spaces and reactor compartment and several people crawled under the casing sniffing like bloodhounds for an aroma but failed to detect any leaks. However, the vitamin C left the compartment smelling like an orange grove for days after the test.

Concurrent with her construction, *Dreadnought*'s crew were being trained in her operation using a wooden mock up of the nuclear submarine at a shipyard in Southampton. Apparently this mock up was slightly thinner than the real submarine and this caused some criticism at the time.

The day came in 1960 for the official launching of Britain's first nuclear submarine and the date chosen for the launch could not have been more appropriate, 21st October, Trafalgar Day. One hundred and fifty five years earlier Lord Horatio Nelson had defeated the combined French and Spanish fleets at the Battle of Trafalgar and on the anniversary of his victory the latest Royal Navy warship would slide gently into the water. Naturally, HM Queen Elizabeth II had been asked to perform the launching ceremony.

The Queen arrived by train at Barrow Railway Station. She was driven through the town taking in Abbey Road, Duke Street and Michaelson Road to the Main Entrance of the General Office at Vickers Barrow in Furness Shipyard. Accompanied by her husband the Duke of Edinburgh, and her party the Queen was introduced to the Lord Lieutenant of Lancashire who in turn presented the Viscount Knollys, Chairman of Vickers Ltd. She was also introduced to the Chairman of Vickers Armstrong Ltd, Major General Sir Charles Dunphie and also the Managing Director of Vickers Armstrong (Shipbuilding), Mr G H Houlder. The submarine had been prepared by shipyard workers and the runs had been greased to allow her to move gracefully into her natural environment. Her Majesty Queen Elizabeth II climbed the stairs to the launching platform and was joined there by the Duke of Edinburgh, Lord Carrington as the First Lord of the Admiralty, Admiral of the Fleet Earl Mountbatten, Chief of the Defence Staff; Admiral Sir Caspar John, the First Sea Lord and other members of the Admiralty. Also in attendance were Commander Samborne and Admiral Hyman Rickover from the United States Navy whose help and assistance had allowed *Dreadnought* to be built within time and budget. A simple religious service was conducted by the Reverend A C Wade, Vicar of St Johns, Barrow in Furness.

HM Queen Elizabeth II had launched a great many warships but *Dreadnought* had received a great deal more media attention than many of her previous launchings. She was after all revolutionary in almost all aspects of her design. The Queen addressed the assembled dignitaries and crowds. "This *Dreadnought* with her great submerged speed and remarkable endurance is worthy of her famous predecessor. The United States has allowed us to draw unreservedly on their resources and on the experience they have already gained with their own nuclear submarines. These are the actions of a true and trusted ally." At 10.50 she spoke the words, "I name this ship *Dreadnought*. May God bless her and all who sail in her." She pressed the lever that indicated full astern which, in turn, released the traditional bottle of empire wine that crashed across the bows of *Dreadnought* to send her on her way. The assembled crowds of visitors, shipyard workers, dignitaries and press cheered loudly as the submarine elegantly slid away from the Queen. The band struck up a rendition of Britannia Rule the Waves, followed by the National Anthem.

The event had been recorded by numerous live television cameras from both the BBC and ITC companies along with a great many journalists from radio and newspapers. These members of the media followed the Queen as she was given a guided tour of the brand new P&O cruise liner *Oriana* which was almost complete nearby to the *Dreadnought*.

At a luncheon organised for the Royal party and invited quests Lord Carrington spoke of the new submarine. "This is a very special occasion for the Royal Navy, which I have the honour to represent, for today we have seen the birth not just of a new ship but of a new era..."

On the last day of February 1962 *Dreadnought* was the scene of a security alert. A group of students, reportedly from Manchester University gained access to the shipyard and to the *Dreadnought* at between 7.30 and 8.30 in the evening. These students then proceeded to daub 'RAG' onto the submarine's hull in a reference to the Rag week stunt they had just pulled off. The Admiralty, however, did not see the funny side of this venture and ordered a full security probe into the incident.

Dreadnought still needed a great deal of attention before she was complete and she was taken to Ramsden Dock where work on her continued for the next two and a half years. The all important loading of the reactor core began in the summer of 1962 and the reactor went critical later that year.

With the power plant on line the ship's company and shipyard workers from Vickers commenced a series of sea trials. In December 1962 *Dreadnought* took to sea for her Preliminary Sea Trials. They were not a complete success. The boat had reached the lock at the basin entrance when the internal communication system completely failed. The fault was traced to a small 3 amp fuse in a sub unit.

When she did get to sea for her 25 hour surface trials Archie McDonald, the Trinity House pilot was onboard to help navigate the submarine. On 13 December 1962 *Dreadnought* returned to the confines of the Vickers Shipyard. Spectators who saw the arrival of the sinister black submarine saw her make her way between the islands of Roa and Piel at about 11.40 in the morning. *Dreadnought* was eased into the Walney Channel by the tugs Roa and Hazelgarth. *Dreadnought* was secured alongside Devonshire Dock by 2.15 in the afternoon.

The winter of early 1963 was one of the coldest in recorded history. On January 9th 1963 in Ramsden Dock *Dreadnought* was poised on the edge of her first dive. Everyone on board was excited and watched dials and instruments carefully. The order was given to flood the main ballast tanks and water entered in. However, *Dreadnought* stayed afloat on the surface despite the flooded ballast tanks. The reason was found to be that some calculations had been done incorrectly. It was felt that the tests should be abandoned for the day and that the next day with new calculations they would have another go. The next day a crane heaped extra ballast on the casing and the diving drill was repeated. Cheers were heard on board as *Dreadnought* settled slowly to the bottom of the Dock with her fin visible pointing straight out of the water that was surrounded on all sides by layers of wintry ice and snow.

Dreadnought was in February 1963 anchored in Rothesay Bay conducting further trials and tests. The submarine received word that the weather was starting to deteriorate and the Captain decided to make for safe waters. At that particular time the nuclear reactor had been shut down and the submarine was running on her auxiliary diesel generator. The personnel in charge of the nuclear reactor saw to it that it went critical in short measure and the *Dreadnought* was taken to safer waters.

Later in her diving trials, Professor Douglas Faulkner recalls some interesting stories about the Russian's interest in the latest Royal Navy submarine. "The Russians, of course, were very interested to know what our speed and deep diving depth were. Their trawler, which always met our submarines at the mouth of the River Clyde on their deep diving trials, radioed our accompanying frigate to ask where the big fish was going. The reply given was Greenland. Ivan was unconvinced, so we played a hedge hopping game of gradually increasing speed until we were beyond detection, and then doing our full speed and deep dive trials in stages until the trawler caught up."

"During one of the dives the 'O ring' seals around the shaft which raised and lowered the snort mast blew inboard (the retaining ring having been secured incorrectly) followed by a strong water jet. I was about halfway through the normal 10 minute cycle of logging the strains on the Kelvin Hughes equipment, when the Captain enquired how much longer was needed. I said 5 minutes, and was impressed when he replied he would try to hold trim and depth steady for the remaining readings before coming up to remedy the cause."

These trials continued almost until the time of *Dreadnought*'s official handing over to the Royal Navy. The trials of the first of any new class of warship are very extensive and such was the case with *Dreadnought*, perhaps more so as she was so revolutionary.

The post Contractors Sea Trials conference was held on 12 February 1963 and highlighted any areas of concern. The Admiralty Engineer Overseer had presented 193 points, which needed some, albeit minor, attention. These ranged from Perspex guard to be fitted over stern seal to prevent splash, stainless steel cladding to be laid over main turbines to ahead manoeuvring throttle linkage stiff at depth. The conference itself assessed each 29 different trials and most were deemed satisfactory. There were, of course, exceptions - the measured mile run was not completed due to poor visibility on the day chosen. Her periscope and mast trials showed the masts to be slow due to defective valves. New piping was fitted to both periscopes and the AKU mast, both of which vibrated. The gash ejector trials were also deemed to be unsatisfactory as the muzzle door did not seat properly. During one phase of the trials it was found to be stiff in operation and indeed during one full speed submerged trial run a probe broke free from the submarine. During the depth charging and steering manoeuvres phase of the trials *Dreadnought* gave the 'impression of less stability than *Skipjack*. Possible requirement to increase area of fixed control surfaces.' Perhaps one of the most important elements of a submarine is that she should be quiet, but *Dreadnought* wanted to be heard or so it seemed because at the conference the delegates were told that *Dreadnought*'s propeller sang. This required a new quiet propeller to be fitted to the submarine at the earliest opportunity. Another problem that was to plague *Dreadnought* in the years ahead was that her hull vibrated at critical speeds when submerged.

Upon *Dreadnought* leaving Barrow in Furness for the last time on St George's Day 1963, Vickers Armstrong Shipbuilders band played as the submarine slowly slipped away from the yard. The tunes chosen were 'Anchors away, Life on the Ocean Wave and Rule Britannia.'

Chapter Two

First Commission, 1963 -1968

Dreadnought commissioned into the Royal Navy on 17th April 1963 when the commissioning cake was cut. During the ceremony the ship's commanding officer and his men stood bare headed in heavy rain. At the time *Dreadnought* was seen as the future of the Royal Navy and represented an epochal transition to nuclear power from diesel and steam. One fact remained; Britain had joined a very select group of nations who possessed nuclear submarines; just America and Russia before the United Kingdom, although in the years to follow, France would soon join the club. Despite the difficulties of her trials, *Dreadnought* was completed on time and within the budget and became the only post war defence project to do so, until the Polaris programme was completed. Even so *Dreadnought* was an expensive investment for the Royal Navy costing £18,455,000 at 1963 prices. A former submarine commander, the Rev Ian Marchant, conducted the commissioning ceremony. The Chaplain blessed the boat with the words, "Go forth into the world of peace, be of good courage, hold fast that which is good: render no man evil for evil: strengthen the fainthearted: support the weak: help the afflicted: honour all men: love and serve the Lord: rejoicing in the power of the Lord." Finally the Chaplin gave the blessing to the submarine. Commander Samborne ordered 'Sound the Alert' and signalled for the band to play the National Anthem and for the colours to be hoisted. The final part of the ceremony saw the ship's company march off the vessel.

The twelve months after commissioning was a period of extensive trials to prove the submarine and her equipment. *Dreadnought*'s first exercise was entitled Exercise Unison and saw her operating with many new units of the surface fleet including the aircraft carrier *Hermes,* the cruiser *Tiger* and the guided missile destroyer *Hampshire*. On the 11 September 1963 *Dreadnought* was at Faslane. Exercise Lime Jug was a large exercise that saw the co-ordinated use of modern naval power both above and below the waves. Held over eight days at the beginning of December *Dreadnought* was joined by the *Olympus* from the 3rd Submarine Squadron and *Porpoise* from the 1st Submarine Squadron. Above the waves were *Hermes, Hampshire* and ten other surface vessels.

Dreadnought put into Rosyth Dockyard on 14 December 1963 under very tight security. The submarine berthed at No 3 dock and special lapel badges were manufactured that restricted to the bare minimum those allowed to enter or work near to the prototype nuclear submarine. Furthermore, those allowed to board the vessel had been thoroughly vetted by the security services and were considered not to pose a security risk. One crewmember recalls that at night in Rosyth the submarine was bathed in the light of a number of arc lights as another precaution against possible sabotage. As this was the first of many visits *Dreadnought* would make over the years to Rosyth the dockyard commander, Rear Admiral J G Watson organised a press conference for local and national journalists on board the *Dreadnought*. He addressed the members of the press and said, "We are particularly happy to have the opportunity of seeing *Dreadnought* and getting conversant with the equipment we will eventually be called upon to refit." The refit of which the Admiral spoke was some years away but planning had already commenced for it. Indeed, Rosyth Dockyard was being transformed with the construction of a brand new nuclear submarine refitting complex. The Admiral continued to explain to the members of the press assembled onboard the submarine that, "We have every confidence that we can produce a good ship to the satisfaction of the Admiralty and the commanding officer."

On 6 May 1964 she sailed from Faslane for sound trials which were conducted over the following three days. The 10 to the 12 of

May were spent on the surface in the Irish Sea as *Dreadnought* made her way to Portsmouth. The submarine stayed at Portsmouth for four days during which time the C in C Portsmouth paid a visit and was escorted around the submarine.

The new torpedo systems onboard *Dreadnought* needed extensive testing and as part of this process the submarine was dispatched to the Mediterranean in the summer of 1964. For almost a month between 20 May and 12 June she was based at Gibraltar from where the torpedo discharge trials were conducted and evaluated. Some of the Mk8 and Mk20 torpedoes had a degree of trouble leaving the submarine, which prompted some redesign of the torpedo release system. The torpedoes fired were inert test weapons and were ballasted instead of containing live warheads. Even so a number of these weapons took a considerable amount of time to retrieve from the ranges.

The following month saw another VIP visit to the submarine in the form of the First Sea Lord, Sir David Luce and his Naval Assistant Captain (later Second Sea Lord) David Williams come onboard the submarine. They toured the *Dreadnought* for a number of hours and were both said to be very impressed with the submarine and her crew.

Dreadnought left Gibraltar on 12 June and set a course across the North Atlantic to visit the US Navy Base at Norfolk, Virginia. The dangers of operating submarines are always at the back of submariners minds. Rarely, however, do these fears come to the fore. The crew of *Dreadnought* could, however, have been forgiven when during 'ship handling exercises' in the North Atlantic, drama occurred onboard. The boat was in a diving turn when suddenly she refused to answer the planes. The depth kept increasing rapidly as the nuclear submarine continued to head for the bottom of the sea. The rudder was acting as a diving hydroplane when turning at high speed. Control was re-established when the Captain ordered the engines be put astern. The submarine recovered and levelled out and an interesting record was most probably placed in the Captain's log about *Dreadnought*'s diving characteristics.

Despite this incident the voyage across the North Atlantic was fast enough to break the record for a nuclear powered submarine. On 19 June 1964 *Dreadnought* secured alongside the submarine tender USS *Orion* for some nuclear work at Norfolk, Virginia. The *Dreadnought* was also very close to one of the latest American nuclear submarines USS *Triton*. This vessel had two reactors and access was extremely strict with many disappointed members of the *Dreadnought*'s crew being denied access. The American's were eager to see that 'their' nuclear reactor was being maintained properly and were keen to make any necessary adjustments. The *Dreadnought* had been invited to Norfolk by the crew of USS *Shark* and amongst many enjoyable activities the American's laid on for the *Dreadnought*'s crew were visits to Williamsburg, beach barbecues and many dances. Five days later *Dreadnought* headed back out to sea for a VIP cruise to the Vacopes area. *Dreadnought* sailed on 24 June for a 36 hour exercise with USS *Dace* and USS *Hardhead*. In the first phase of the exercise, the USS *Hardhead* acted as a snorting transitter. USS *Dace,* meanwhile operated as an attack submarine. *Dreadnought*'s role was to play an intruder. The second phase was a leapfrog SSK versus transitter with USS *Dace* and *Dreadnought* alternating roles. *Hardhead* was withdrawn in view of the Flag Officer Submarines' directive not to allow a direct comparison between *Dreadnought*'s Type 2001 sonar and the American's BQS6 mounted on the USS *Dace*. In the end neither the American or British boat would successfully complete an attack sequence. A two-day exercise with elements of the US Navy happened between 24 and 26 June whilst more trials were conducted whilst on underwater passage to Bermuda.

Dreadnought made an excellent display in the Caribbean sunshine at Bermuda and amongst many other VIP's welcomed onboard was the Governor of the island. The *Dreadnought* had arrived in late June 1964 just after the completion of that year's Tall

Ships race and the harbour was awash with classic multi-masted vessels mostly fully rigged and flagged. Only three days were spent on the island paradise before *Dreadnought* would transit the North Atlantic again this time headed in the direction of Gibraltar for more torpedo trials.

John Howe recalls an incident during these trials. "The Governor of Gibraltar was on board *Dreadnought* on a day when torpedo firing trials were being carried out. The practice torpedoes used surfaced after a set run distance and were recovered by a motor fishing boat (MFV) which, on this particular day, was in the charge of a seconded officer from the *Dreadnought*.

The forenoon trials went as scheduled but conscious of the Governor's liking for a tot of navy rum, the Captain surfaced *Dreadnought* immediately prior to mid day and gave the order 'Up Spirits', a ceremonial that was conducted by the coxswain and a designated officer. When the keys to the Rum Store could not be found, the officer of the MFV was contacted by radio and confirmed the appropriate keys were in his pocket.

The Captain immediately ordered the MFV to come alongside *Dreadnought;* after a few aborted attempts a rather irritable Captain, under the gaze of a thirsty Governor, told the Officer in Charge of the MFV to 'Heave to' and he would then bring *Dreadnought* alongside him.

So, rather belatedly, the Governor and the Ships Company, downed their tots with the traditional 'The Queen, God Bless Her' toast. The torpedo trials lasted until 4 August when *Dreadnought* headed back to Portsmouth. After a day stop in Portsmouth *Dreadnought* sailed to Faslane arriving there after a two day passage on 10 August. Almost a month was spent alongside for leave and essential maintenance.

At midday on 17 August 1964 Commander Barnaby Samborne's time in command came to an end when he was relieved by an officer who would become one of the most distinguished military leaders since the Second World War, Commander John D E Fieldhouse (who would later become Admiral of the Fleet Lord Fieldhouse). Commander Fieldhouse entered the Royal Navy at Dartmouth in 1941 and would later serve as a midshipman on the East Indies Station. In 1948 he started his submarine training at the home of Royal Navy submarines, HMS *Dolphin* at Gosport, before becoming 3rd and 4th Hand and 1st Lieutenant of four in the 3rd Squadron's Submarines based at Rothesay. He had risen up the ranks until in 1956 after completing the Commanding Officers Qualifying Course (COQC) he was given command of the 'A' class submarine *Acheron*. Two years later in 1958 he commanded the *Tiptoe* and later the more modern *Walrus* between 1961 and 1963 before being appointed as commanding officer of *Dreadnought*.

In mid September 1964 *Dreadnought* was worked up in the Clyde area and each night the submarine would anchor at Lamlash. 21 September saw the start of Exercise Teamwork which would occupy the crew through until 2 October. At its height this exercise saw no less than ten submarines, 25 minesweepers, one minehunter, one minelayer, two coastal store carriers, one escort vessel and seven frigates taking part. At its completion *Dreadnought* returned to Faslane where further sonar trials were conducted in the Clyde. Another eight-day exercise took place from the 13 October before a trip south to Portsmouth where she spent four days. Crew members took the opportunity of a run ashore in Portsmouth before returning to the submarine for another three day exercise and thence to Faslane on 1 November. The cold winter in Scotland was left behind as the submarine headed southwards and for Gibraltar. During the passage yet more exercises were conducted before the nuclear submarine put into Gibraltar on 7 November. Sonar trials were conducted over the following months in and around the waters of Gibraltar and the Mediterranean, but *Dreadnought* never ventured too far from Gibraltar because she returned to the port during the weekends when leave was granted to the crew. On 5 December *Dreadnought* headed for home and returned to Faslane 11 December.

Late February 1965 saw *Dreadnought* alongside at Rosyth for docking and leave. During the passage from Rosyth to Faslane *Dreadnought*'s systems and crew were worked up. Also during the passage around the North of Scotland a visit was organised to the nuclear reactor site at Dounreay. *Dreadnought* reached Faslane on 27 February and stayed there until 2 March when she sailed for exercises in the Londonderry area. These were quickly followed by exercises in the Clyde between 8 and 12 March. The following day the submarine returned to Faslane for seven days leave and maintenance. She would spend the best part of April in the local exercise areas working up and on 26 April was inspected by Captain S/M 3's Inspection Leave. The sea inspection being conducted in waters around Londonderry.

29 April to 12 May were spent conducting Exercise Red Knight with *Ocelot, Olympus, Otus,* and *Opossum* and a surface unit before a five-day stop back at base in Faslane. The crew used this time to prepare for a three-day official visit to Southampton on 19 May and secured alongside berth 109 in Southampton Docks. The three-day visit saw parties from local organisations visit the submarine and the Mayor of the City Mrs E E Willock, JP visited on Thursday 20 May. The submarine sailed around the Solent and entered Portsmouth harbour on 22 May and stayed there for a further four days. Again the sunnier climes of Gibraltar beckoned the submarine away from a British summer and the Rock was reached on 29 May. Between the 31 May and 10 June *Dreadnought*'s crew were working hard to achieve the objectives for Opeval which was held over two days from the 5 June in Palmas Bay. Further exercises and sea trials kept *Dreadnought* busy until 4 July. The submarine again turned north and headed up the English Channel and towards the Netherlands where she made an official visit to Rotterdam together with the submarine depot ship *Maidstone* between 9 and 14 July. It just so happened that on the 9 July the

Dutch Royal Barge carrying Queen Juliana and Prince Bernhardt was to pass the submarine at midday. The Captain decided to don his best dress uniform and take the salute on the casing of *Dreadnought*. He dressed in a hurry and just managed to get the salute in time. There have subsequently been many rumours that the Captain was quite concerned that he had failed to do his flies up during the salute! This could well be some well-intentioned Michael taking by matelots. *Dreadnought* returned to Portsmouth the following day and received word that she would undertake noise trials on her passage back to Faslane between 19 and 23 July. After a two-day stop in Scotland, Northern Ireland welcomed the submarine back to the waters around Londonderry where more trials were carried out.

Dreadnought was on public display, this time in Scotland and the occasion was a review of the fleet by the Queen. The Review of the Home Fleet in the Clyde had been in the planning for many months and *Dreadnought* was to play a major part in the review. She was positioned on R line in position R2. R1 was taken by the submarine depot ship *Adamant* and the submarines to her other side were in order *Otus, Osiris, Oracle, Walrus,* and *Narwhal*. The Royal Yacht *Britannia* passed *Dreadnought* and the other 72 ships in line on Tuesday 10 August 1965. Once the Royal Yacht had passed *Dreadnought* was moved from her anchorage and past the frigate *Puma* on J line to alongside the submarine depot ship *Maidstone* which occupied position J5. At 1400 Queen Elizabeth boarded *Maidstone*. Whilst onboard she spoke to senior ratings of the Third Submarine Squadron which, of course, included crew from *Dreadnought*. The Queen left *Maidstone* and took a short boat trip to the nuclear submarine she had launched five years earlier. The Royal Barge drew alongside *Dreadnought* to be met by the commanding officer, Commander John Fieldhouse. The *Dreadnought* received onboard a seemingly endless Royal procession. The Queen was accompanied by The Duke of Edinburgh and young Prince Charles and Princess Anne. Admiral of the Fleet Lord Mountbatten of Burma made up the royal party. Behind the Royals came the Admirals which included Admiral Sir David Luce the First Sea Lord, Admiral Sir John Frewen, C in C Home Fleet, Admiral Sir Wilfred Woods, First and Principal ADC to the Queen, Rear Admiral McGeoch, Flag Officer Submarines and Rear Admiral Morgan, Flag Officer Royal Yacht. General Sir Rodney Moore - defence services secretary and the Minister of Defence Mr Christopher Mayhew made up the party.

Queen Elizabeth II toured the submarine calling first at the wardroom to met the officers of *Dreadnought* and also to change her footwear to something more appropriate for touring an operational submarine. Her shoes and handbag were handed over to Petty Officer Steward Fenton who looked after her Majesty's possessions until her departure from the boat.

The Royal Party was shown all around the submarine and was given demonstrations of water shots from the torpedo tubes. The Queen and Prince Philip were shown the accommodation spaces and Senior Rates Mess before passing 44 Bulkhead and entering the 'American sector'. This is the only time that the Containment has been broken on board when the Reactor has been Critical and the plant self sustaining - the tunnel being open at both ends.

During her visit the Queen saw the order given to open Main Vents. This was necessary to comply with Royal protocol. The rule being that a royal step should be no more than 7 inches. The step down from the casing ladder into the Royal Barge was found to be more than this. Main Vents on No 2 Main Ballast Tank were opened to lower the casing the required amount. The Queen and her party were piped off *Dreadnought* and were later joined by Prince Charles and his sister Princess Anne who had been given a separate tour of the submarine.

The Review over, *Dreadnought* sailed on 12 August for Rosyth, a voyage which would take her two days. *Dreadnought* stayed at Rosyth for essential maintenance and a mini refit through until 2 February 1966. If *Dreadnought*'s nuclear core was proof of progress, equipment to monitor such devices was also developed in parallel and 1965 saw the introduction of detailed surveys to check for minute structural failures. This equipment used Ultra Sonic Crack Detection (NDT) equipment and other means to discover hairline cracks that could cause 'problems'. During a survey on *Dreadnought* such hairline cracks were discovered in internal bulkheads. The Ministry of Defence press office were asked by journalists how significant these cracks were and dismissed them as trivial. The spokesman said, "The cracks did not represent a serious structural failure. They were not dangerous to the vessel or her crew, and could be repaired by welding during normal docking." Despite the supposed triviality the cost of repairs to the *Dreadnought* was £150,000 at 1965 prices.

Commander Fieldhouse was relieved in command of *Dreadnought* by Commander Peter Cobb in April 1966. He was born in 1930 and was educated in Canada before joining the Royal Navy at Dartmouth in 1947. He served for a time onboard the ill-fated *Affray* in 1950. He would also serve in HM submarines *Tudor, Artful, Selene, Andrew, Telemachus* and *Tactician*. In 1959 he was appointed as commanding officer of *Seneschal* and then spent two years at HMS *Ganges*. Later he became Staff Officer Operations with 5th Submarine Division at Malta before commanding the *Narwhal*.

Post refit trials continued through to 30 May in British waters especially around Faslane and the Clyde. *Dreadnought* began some trials and exercises, with *Osiris* between 2 and 4 June. Later trials continued around Gibraltar where the submarine stayed for four days between the 8 and 12 June. She then proceeded to Loch Fyne and on to Rosyth. *Dreadnought* stayed at Rosyth for just under a month between 30 June and 25 July 1966 and this period was followed by yet more trials.

Dreadnought was again in dockyard hands for an extended period between 30 July 1966 and 25 January 1967 as workers at Rosyth carried out essential repairs and modifications to the boat.

On two days from 14 February *Dreadnought* conducted torpedo firings in the Clyde before entering Faslane for a couple of days until 19 February. She left Faslane that day and headed for Loch Fyne for a period of seven days through to the end of the month. 1 March saw the submarine heading for Faslane and on the 3 March she left to sail to Rosyth for a leave and maintenance period which occupied almost the whole of March. Once complete *Dreadnought* headed back to Faslane and later in the month of April undertook a twenty-day exercise. During the course of 1967 a serious incident which could have had catastrophic results for *Dreadnought* occurred. The submarine was running at high speed and suffered from a number of leaks in brazed pipe joints. These leaks allowed seawater to be sprayed over the reactor control equipment. *Dreadnought*'s internal design was very similar to the US *Skipjack* class and it has been speculated that a fault of this kind sank USS *Thresher* in 1963. An inquiry into the incident led to revised practices in the fabrication and fitting out of British nuclear submarines.

Dreadnought participated in Exercise Quick Pursuit in the Forth area of Scotland. Also taking part were the submarines *Otter, Porpoise, Grampus, Olympus* and *Truncheon*.

After a brief period at Rosyth the submarine sailed for Gibraltar on 8 June 1967 but the course chosen took the boat around the north of Scotland and also saw a brief stop at Faslane. *Dreadnought* reached Gibraltar on 19 June and entered into a maintenance period and to evaluate data from her sonar. Indeed *Dreadnought* was docked down in the commercial dry dock for what was expected to be a period of ten days. On the 21 June she had much of her crew on leave including the Captain in Tangiers and Senior Technical Officer away on ships business. Much of her sonar was in pieces receiving attention when the duty officer of the boat, Lieutenant J Higgins received an urgent call asking how quickly the *Dreadnought* could be brought into an operational state. His reply was twenty-four hours.

The reason for the enquiry was that the German tanker *Essberger Chemist* had put out an SOS call off the Azores. She was taking in water and adrift off the Azores after her engines had failed. From a docking state to engine ready, full war readiness took the crew of *Dreadnought* just 12 frantic hours.

Being in a civil dry dock the *Dreadnought* was unable to start her nuclear propulsion until she was in the naval part of Gibraltar harbour. She was eased out of the dock and taken to her normal berth on the arm of South Mole. The reactor was brought online and at 10.00 on the 22 June *Dreadnought* sailed.

The Captain received the call at his accommodation in Tangiers and quickly organised his return trip arriving with just 15 minutes to spare before her departure. *Dreadnought* set sail and headed towards the Azores and dived soon after leaving Gibraltar. The 1000 miles was covered in two days at full underwater speed. The submarine surfaced in a calm sea some 15 miles away from the drifting tanker, which had failed to sink of its own accord.

The last remaining crewmembers of the *Essberger Chemist* had been taken off the tanker which had started to drift into the shipping lanes and posed a considerable danger. Accordingly on 24th June 1967, *Dreadnought* received the order to sink the ship. The Captain decided to take his time and plan his course of action. A number of passes of the wreck were made and contact was established with the frigate *Salisbury* and Royal Air Force Shackleton maritime reconnaissance aircraft in the area. The *Essberger Chemist* was drifting at a speed of around one knot and had also developed a heavy list to starboard. The Captain told all other vessels in the area to stand clear including a tug and the *Salisbury*. *Dreadnought* dived and positioned herself about a mile away at action stations. All watertight doors were shut. This was considered necessary because of the tanker's explosive cargo of 6000 tons of acetone and alcohol. Four torpedoes were fired but one missed, 'very close and just astern'. The three torpedoes that successfully struck the tanker hit along the length of the target. The explosions, whilst very impressive on the surface resulted in 'thud' noises onboard *Dreadnought*. The tanker at first appeared to be sinking but she remained stubbornly just afloat on the surface held up by some of her undamaged tanks. The frigate *Salisbury* was called in to finish her off with a number of 4.5" shells. *Dreadnought* surfaced once again and along with the *Salisbury* surveyed the flotsam and jetsam on the surface. All that remained was a burning liquid cargo of alcohol that fuelled a column of smoke thousands of feet high.

Her job done *Dreadnought* returned to Gibraltar on 28 June and re-entered the dry dock to complete the maintenance on her equipment. This was completed on 9 July. There followed a series of exercises in the Gibraltar area and further afield in the North Atlantic to complete which occupied the submarine for the best part of early July. These were over by mid July and the submarine sailed for the United Kingdom in company with *Alliance* arriving at Loch Fyne on 22 July. Between 27 July and 6 August *Dreadnought* entered into Squadron Sports and came out as winner. The next day saw the submarine sail for Rosyth; whilst en route she took part in a series of exercises in Scottish waters.

The Royal Navy was eager to show just how flexible an instrument nuclear power could be. Accordingly the Ministry of Defence decided a long distance underwater voyage for the submarine would be an appropriate demonstration. On 19th September 1967 *Dreadnought* left Rosyth for Singapore via Faslane on a sustained high speed run.

HMS Dreadnought's keel laying ceremony (Royal Navy Submarine Museum)

The launch of HMS Dreadnought (Royal Navy Submarine Museum)

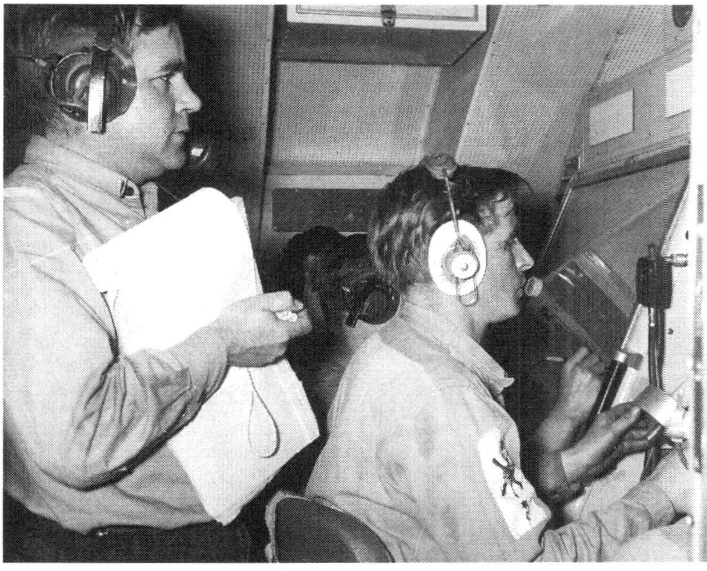

Sonar Training on board HMS Maidstone (HMS Maidstone)

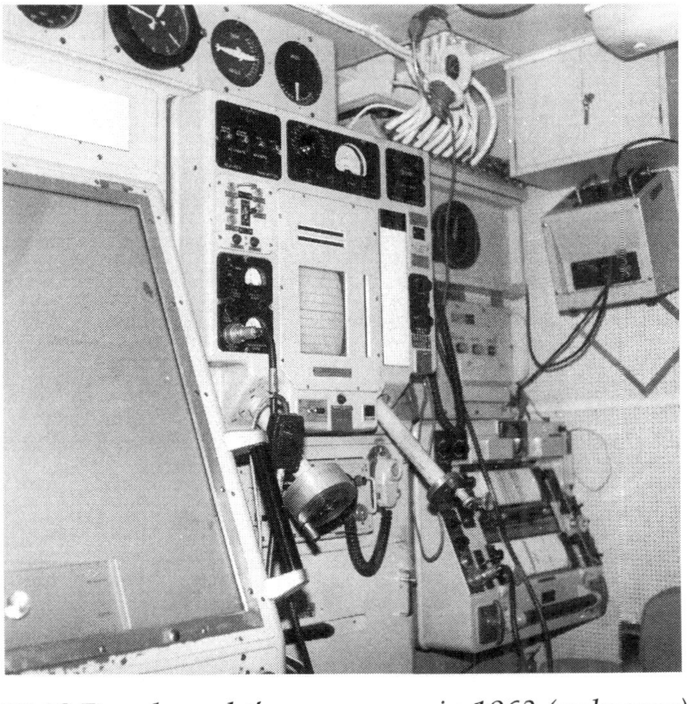

HMS Dreadnought's sonar room in 1963 (unknown)

At sea in 1963 (Royal Navy Submarine Museum)

Two great engineering feats! - Under the Forth Bridge (Royal Navy Submarine Museum)

In heavy seas…. (Royal Navy Submarine Museum)

….and in calm, off Gibraltar (unknown)

Sinking the Essberger Chemist (Royal Navy Submarine Museum)

At the home of HM Submarines, HMS Dolphin (Navpic)

HMS Dreadnought breaking through the ice cap on her North Pole mission (Royal Navy Submarine Museum)

HMS Dreadnought's bow wave at speed (HMS Neptune)

At the Jubilee Spithead Review in 1977 (Navpic)

Dreadnought's sixth commander, Paddy O'Riordan (Crown Copyright)

Entering Chatham for the last time (Royal Navy Submarine Museum)

HMS Dreadnought paying off (Royal Navy Submarine Museum)

After two days in Faslane to load a full outfit of torpedoes she sailed for trial torpedo firings in the Clyde area. These completed, *Dreadnought* started her journey to the Far East. At the 100 fathom line off Northern Ireland *Dreadnought* dived for the fast, deep transit to Gibraltar which was reached on 26 September. Just one night was spent at Gibraltar before the submarine headed once more to sea.

One aspect of nuclear powered submarine patrols that quickly became apparent was the need to relieve the boredom of long cruises. One solution tried onboard *Dreadnought* was the production of a daily paper, the Dreadnought Express. The paper reported news and activities onboard the submarine and also jovially poked fun at whoever deserved some rib taking.

With Gibraltar behind them next stop for the crew would be Langkawi off the west coast of Malaya. Once clear of the Straits the submarine dived and headed for a planned rendezvous with the aircraft carrier *Hermes*. *Hermes* was met off the west coast of Africa for a series of anti submarine exercises. *Dreadnought* acted as the target and by all accounts gave a very good account of herself.

On completion of the exercises the boat broke the surface and transferred *Hermes'* mail to the aircraft carrier. Interestingly *Dreadnought*'s former CO, Commander Fieldhouse, was onboard *Hermes* as her Executive Officer. He arranged for fresh baked rolls to be sent over to the submarine.

On September 29 the lights of Tenerife and Grand Canary were witnessed through the periscope. The following day saw one of the submarine's carbon monoxide burners begin to break down. These so called scrubbers remove harmful CO-H2 build up in the submarine. Luckily *Dreadnought* was fitted with a number of such devices and the faulty one was soon repaired.

At 06.30 on 2 October 1967, *Dreadnought* crossed the Equator and surfaced to allow King Neptune and his Court to conduct the traditional crossing the line ceremonies. Half a ton of accumulated dry gash was also disposed over the side. The opportunity was also taken for some shooting practice. Boxes and drums were thrown over the side to become targets in an Inter-Mess shooting competition. The Senior Rates' Mess won with 16 hits against 15 by the Junior Rates and Wardroom. The ceremonies and shooting match completed *Dreadnought* dived to continue her voyage.

By 6 October, rough weather caused the planned surface passage across the Agulhas Bank, designed to save time, to be cancelled. Instead the submarine headed around the Cape and raised the periscope to check her position with the Cape Light. *Dreadnought* continued underwater passing Madagascar until she was some 30 miles off Langkawi island when she surfaced at night on 16 October. Most of *Dreadnought*'s crew took the opportunity to go on the bridge and were hit by the hot humid atmosphere. They saw a large number of islands dotted on the horizon and the sea was calm.

At Langkawi the submarine joined up with the frigate *Minerva* and submarine *Anchorite* for an exercise that also involved the Australian HMAS *Yarra* and *Stuart*.

Malcolm Perkins remembers how he came to be onboard *Dreadnought* by accident at this time. "I was based in the Far East as a Radio Supervisor. As part of my role on Flag Officers' staff in Singapore, I had to visit ships and check on their standard of equipment training in the Radio Department. My time in *Dreadnought* resulted from a three month posting to an Aussie ship, HMAS *Stuart*. All was well for the first month but there was a change of plans when she was sent at short notice to Vietnam. To avoid political embarrassment it was decided I should leave *Stuart* at the earliest opportunity. The nearest Royal Navy ship was *Dreadnought* which was a short distance away having just completed the journey to the Far East. I was to be transferred by inflatable dinghy with mail for *Dreadnought* which had been dropped to *Stuart* earlier. The sea was unsettled and as I approached *Dreadnought* I had to jump on the upward roll as we say. I had the mail and a small canvas bag with my belongings in my left hand and as I jumped it all went wrong. The wave was not as high as I anticipated and I was deposited on the Med ladder with the iron work just out of reach and began sliding down the side towards the sea. The crew on the deck of *Dreadnought* saw what was happening and one of the seamen lay on the deck and reached towards me and grabbed the mail sack and my bag, leaving me to go overboard where I was retrieved by the crew of the inflatable. I suppose it's all about priorities - if you've been at sea for some weeks and the choice is put before you, I know which one I'd choose. I am not 100 percent sure, but as the seaman grabbed the mail and I started my slide, I'm sure there was a cheer from those on the deck of *Dreadnought*. I made it to *Dreadnought* on my second attempt and had a most enjoyable few weeks while she exercised off Singapore. The way the crew handled her was fantastic - she appeared to be more like a plane than a boat. She did everything except loop the loop when we were taking avoiding action away from practice Ikara launched torpedoes which the Aussies claimed could catch any submarine. Their torpedoes never did catch the *Dreadnought* mainly because of the skill of the Captain and crew. Because accommodation was so tight, I slept in a 'coffin' strapped on the back of a 21" torpedo in the forward section. At night as I lay in my coffin I could look straight up the escape hatch."

The exercises were a complete success and once again *Dreadnought* and her crew gave another good account of themselves. Once the exercises were over members of the submarine's crew took the opportunity of enjoying themselves on the many miles of clear beach for a banyan.

One of the key parts of the exercise was to see how well *Dreadnought* operated in different water conditions, such as varying degrees of salinity and water depths. The crew of *Dreadnought* also took advantage of the hospitality of the foreign ships in the group.

A number of *Dreadnought*'s personnel enjoyed daylong exchanges with their Australian counterparts.

The exercise complete and crew back onboard the submarine, *Dreadnought* headed in the direction of her next stop, the Royal Navy base at Singapore which was reached on 3 November. As Britain's prototype submarine sailed into port she was welcomed by a guard and band reception on the quarterdeck of the submarine tender *Forth*.

For many members of the crew it was the first time they had enjoyed the varied pleasures of the seaport city. Everyone from the submarine was accommodated in HMS *Terror* or as it was known, "the Butlins of the Far East". Generally no one had any complaints with food or messing arrangements but some complaints were made concerning the Chinese laundry and how they over starched crewmembers underwear!

After 17 days at Singapore *Dreadnought* departed Malaysia on 20 November and headed back home to the colder climes of the United Kingdom. The voyage would take the submarine 25 days and would involve retracing her outbound voyage. On 28 November *Dreadnought* surfaced to repair one of the Main Turbine Throttles which had developed a fault. It had been hoped to allow the crew onto the casing but the weather was considered to be too rough to risk this.

On 1 December *Dreadnought* was due to exercise off the Cape of Good Hope with ships of the South African Navy, however, this was cancelled and replaced with a short exercise with a local maritime patrol aircraft. Once the exercise was completed the aircraft dropped a container into the sea, which was eagerly retrieved as it contained a great deal of mail from home.

Seven days later the submarine was heading up the west coast of Africa and rendezvoused with the County class destroyer *Devonshire* off Dakar for a series of exercises before resuming her journey home. A planned ASW exercise on 11 December had to be cancelled at relatively short notice due to the frigate scheduled for the task, the *Euryalus* being firmly in dry dock in Gibraltar. The Royal Air Force, however, took the opportunity to conduct a somewhat ad-hoc exercise with *Dreadnought*. The crew also received more air dropped mail. Off Western Ireland the frigate *Aurora* joined the submarine on her way from Portland and more exercises were conducted. This was followed by another exercise with Royal Air Force aircraft off Northern Ireland.

Dreadnought arrived in the Clyde to fire some torpedoes in pre-arranged trials and evaluations, but these had to be cancelled due to the inclement weather conditions in the area. The nuclear submarine finally arrived at Faslane and tied up on 15 December. Two days were spent at Faslane before the submarine sailed around the coastline of Scotland on the surface to position herself at Rosyth which was reached on 21 December. Christmas and New Year leave was granted which saw *Dreadnought* in dockyard hands for the whole of January and into the early part of February.

Key personnel returned to the boat, which was made ready for sea on 6 February. Between the 7 and 8 February 1968 *Dreadnought* was in transit to Faslane after her time at Rosyth. She arrived at Faslane on the 9 February for an all too short stay. On 12 February she sailed again this time heading for an exercise at sea which was conducted en route to an official visit to the German city of Kiel on the 16th. *Dreadnought*'s berth was alongside Tirpitzmole on a bitterly cold and icy day. The officers and men of *Dreadnought* welcomed onboard a seemingly endless stream of dignitaries and visitors who toured the submarine. Amongst these were high-ranking officers from the German Navy. There was much interest in the German Nuclear Merchant vessel *Otto Hahn* which just happened to be fitting out in the port at the time. *Dreadnought*'s crew also took the opportunity to explore and examine the latest additions to the German Navy, which included a comparatively small but still very capable 300 ton U-boat and the training ship

Ruhr. *Dreadnought*'s crew were given shore leave and they enjoyed trips to the towns of Lubeck and Hamburg with its world famous breweries. Another trip was put on that took in the city of Berlin. The organisation of the visit must have been complicated as it involved flying into the then besieged city and a tour was laid on of all the cities highlights including the Berlin Wall, the 1936 Olympic Stadium, the Brandenburg Gate and the real Checkpoint Charlie.

Having left Kiel *Dreadnought* proceeded to sea and conducted some exercises before heading south towards the sunshine of the Mediterranean. Gibraltar was the next port of call on 3 March where she would stay for the next five days. More exercises was conducted as she headed home arriving at Faslane on 16 March. More torpedoes were loaded and Captain S/M and Staff joined the submarine. She sailed on St Patrick's Day for Sea Inspection, which was conducted over the next three days. On 21 March *Dreadnought* sailed into Faslane whilst on the 22 and 23 March the crew prepared the boat for another Harbour Inspection. *Dreadnought* and crew passed the inspection and sailed for Rosyth on 29 March.

Dreadnought's nuclear reactor core had been starting to show that it needed replacement and plans had already been put in hand to redevelop an area of Rosyth Dockyard in Scotland for the role of Submarine Nuclear Refit. The number of such boats was set to increase steadily throughout the 1970's and such shore-based facilities would become increasingly important. By June 1967 the extensive facilities were ready for their first customer.

Dreadnought's crew prepared the vessel for Britain's first ever nuclear submarine refuelling and refit. Politicians at Westminster watched developments at Rosyth and on *Dreadnought* with a great deal of attention as the programme had already cost many millions of taxpayer's money and an embarrassing situation was to be avoided at all costs.

Dreadnought sailed for Rosyth and headed around Scotland on the surface. Early in the morning of 31 March 1968, forty-five relatives, some reporters from the local newspapers and a BBC camera team joined the submarine by tug at Rosyth and rendezvoused with *Dreadnought* east of the Forth Railway Bridge. The visitors were greeted warmly by all onboard and once safely aboard the submarine steamed to a point off May Island and dived under the surface. The families of the crew were fascinated by the experience but it was over all too quickly for some as the need to catch the tide for entry into Rosyth Dockyard meant *Dreadnought* only remained submerged for an hour. After surfacing she steamed back to Rosyth arriving at 1600.

Dreadnought's first commission ended on 30th April 1968.

Chapter Three

Second Commission, 1969-1974

Whilst *Dreadnought* was in the hands of the skilled workforce at Rosyth Dockyard the local MP for Gosport and Fareham, a Dr Reginald Bennett, courted controversy with the Royal Navy when in the House of Commons he said of *Dreadnought*. "If she is just going from one refit to another, this may give substance to what I have been hearing that she is not a particularly effective ship." He went on to ask if the submarine was "fully seaworthy, capable of proper operational use, and ready for the deep diving for which she had been designed."

On 6 January 1969 Commander Alan Gilbert Kennedy took command of *Dreadnought*. He was born in 1934 and was a native of Woolwich in London. In 1950 Commander Kennedy attended Dartmouth Naval College and went on to serve onboard the A class submarine *Acheron*. Alan Kennedy served as *Dreadnought*'s first Navigating Officer and then joined as the executive officer of the Strategic Missile Submarine *Repulse,* before becoming *Dreadnought*'s commanding officer.

Dreadnought would spend the next two years in dockyard hands at Rosyth in Scotland where her reactor cores were replaced and the inside and outside of the hull was completely scrubbed to remove both natural and artificial contaminants. During the refit process thousands of people worked on this multi million pound project from professions as diverse as calibration testing, tug masters, x-ray technicians, lagging machinists, electricians and those responsible for high duty welding. During the refuelling process plastic sheets were used to cover and completely wrap equipment and surfaces to keep it absolutely dust free. This zero dust environment was absolutely vital to ensure the success of the refuelling process. To further comply with these demanding standards all personnel entering the Refit Complex were ordered to don protective clean clothing, the so called 'bunny suits' for working within the submarines reactor compartment. *Dreadnought*'s first refit, however, ran into a number of problems. There were limited nuclear engineering personnel and other problems concerned the organisation of the dockyard. During the refit it was discovered that there was a serious corrosion problem on many of the submarine's pipes. This had been caused by a compound eating at the metal. The compound itself was formed by water reacting with a substance found in the original lagging used on the pipes. A new form of lagging was installed which didn't react with water in the same way. This did, however, mean the completion date of the refuelling and refit would be put back by a couple of months. During the refit many changes were made to the internal structures in *Dreadnought* but some other additions were made to improve her abilities. *Dreadnought* had always been something of a noisy boat underwater and engineers changed *Dreadnought*'s ballast tank valves in an attempt to reduce some of this self generated noise. She re-commissioned on 10th September 1970 in a religious ceremony conducted by the Reverend A A Upton from HMS *Lochinvar*. *Dreadnought* then commenced a long series of trials and operations to achieve full operability.

It wasn't, however, going to be plain sailing for Britain's prototype submarine because a few hours before she was to sail on 28 September her departure had to be postponed due to a leak in one of her air systems. Engineers decided that due to the inaccessibility of the area where the leak was it was best to deal with the problem in port and the one-day delay was enforced.

Anglo American joint operations throughout 1970's sought to organise intelligence gathering missions in the Barents Sea. The Americans, with more nuclear submarines, took on the majority of the missions, which were given the codename Holystone. With only a handful of operational nuclear boats Britain could only offer to undertake one spring and one autumn mission. One of these missions was undertaken by *Dreadnought* whilst the other was completed by *Valiant*.

2 October 1970 *Dreadnought* did eventually leave Rosyth under her own power and proceeded to the Clyde and Forth areas to conduct numerous trials. Between 6 and 13 October the submarine operated around the Clyde estuary before putting into Faslane the following day. The trials continued for most of the following fortnight until 24 October when she was alongside at Faslane for repairs and maintenance. *Dreadnought* left her berth on 5 November and headed for Loch Striven where she stayed between 6 and 14 of the month before returning once more to Faslane.

She operated in and around the Clyde for the next few days through until the end of November. The new year saw the submarine participating in Exercise Creaky Nut off Lerwick between 11 and 17 February. At the completion of these exercises *Dreadnought* returned home once more to Faslane and operated in the area for the next few days.

It was during the last year of her refit that the Ministry of Defence decided to follow the lead of the United States Navy in having a nuclear submarine surface at the North Pole. The Americans achieved the feat first with USS *Nautilus,* the world's first nuclear powered submarine. Britain was going to do the same but with Britain's first nuclear submarine, *Dreadnought*. There were operationally sound reasons for sending a nuclear boat on such a hazardous mission. Russian Ballistic Missile Submarines would use the Arctic Circle as the ideal launching point for their missiles and this area is the traditional submarine hunting ground. For this mission *Dreadnought* would be accompanied by the diesel electric boat *Oracle* to act as backup. Commander Alan G Kennedy took *Dreadnought* out of Faslane and sailed north with a civilian scientist

Dr Charles Swithinbank from the Scott Polar Research Institute onboard. Upon reaching the outer limits of the ice pack exact navigation was required as communications would be extremely difficult. The task ahead for the crew of *Dreadnought* was considerable, navigate underwater and under ice for around 1500 miles over seven days to reach the true North Pole, break through the ice cap, take scientific measurements and then submerge again and head home.

Dreadnought crossed the Arctic Circle on 22 February and this saw King Neptune and his court come into being. Novices aboard the submarine that had not crossed this line before were 'initiated' into the 'honourable company of Bluenoses'. Crewmembers remember a great deal of good humour and above all else a lot of very cold water into which they were dunked without too much ceremony!

Dreadnought surfaced six times during the voyage to the Pole but finally on 3 March 1971 she became Britain's first submarine to surface at the North Pole. The exact location of the North Pole was reached at 8 am but a suitable place to break through the ice nearby had to be found. The *Dreadnought* eventually found a suitable location and started to smash her way through the Arctic ice. The submarine broke through pack ice around one foot thick after several attempts to crack the ice at 5.30 in the afternoon. No one was under any misconceptions about the cold but even still some of the crew were taken aback when the measurements were taken and the ambient temperature was recorded as minus 37 degrees Fahrenheit. The scientific equipment was quickly deployed and samples taken whilst a watch was kept from the top of the conning tower for passing polar bears. The man at the top of the conning tower was armed with a powerful rifle to shoot the bears if they came too close. The passage out of the ice pack was as demanding as going in, so when the edge of the ice pack was reached it was with some relief. *Dreadnought* returned to warmer climates at Faslane on March 11th 1971. Upon the submarine's return to

Faslane, Commander Kennedy lifted a block of ice from inside *Dreadnought* and threw it off the edge of the jetty at Faslane into the loch. "I think it should float" Commander Kennedy said and indeed the ice floated for a while before melting. At a press conference held after the submarine's return Commander Kennedy fielded a great many questions about the submarine, its performance and how the crew coped with the extreme conditions at the North Pole. Commander Kennedy said, "It has been of great value in that we have learnt a lot about operating in a hostile environment. We went at the worst possible time from the point of view of ice, temperature and daylight and yet there are only a few dents." A number of firsts were achieved during the mission to the North Pole. These included the first Briton to reach the North Pole from underneath which was awarded to Leading Seaman William Brown whose bunk was the furthest forward in the boat and was therefore the first to reach the Pole. Unfortunately for Brown he was asleep at the time of his great achievement. A 19-year-old crewmember became the youngest Briton to walk at the North Pole and one sailor celebrated his birthday at the Pole. Commander Kennedy's achievement in getting *Dreadnought* safely to the North Pole and back was recognised later in the year when at a glittering ceremony held at London's Savoy Hotel he was named 'Man of the Year' and received his prize from the then Foreign Secretary Sir Alec Douglas-Home.

Commander Alan Kennedy's tenure as commanding officer of *Dreadnought* came to an end on 7 June 1971 and he was given a desk job at the Ministry of Defence. To reflect this, he was towed from his command by his crew on a mock desk with wheels. On 7 June 1971 command of *Dreadnought* was passed to Commander Graham C.B Rogers.

Commander Graham Rogers joined the Royal Navy as a cadet in 1949 and passed through Dartmouth Naval College before joining the training cruiser *Devonshire* in 1951. He later served on the aircraft carrier *Theseus* and the frigate *Tintagel Castle* before joining the submarine service in 1955. His initial appointments were to the S class submarine *Sanguine* and the T class *Trenchant* both operating in the Mediterranean. 1957 saw him transfer to *Turpin* at Gosport as first lieutenant. Two and a half years later he joined her sister ship *Teredo* this time in Far East waters.

Commander Rogers passed his submarine commander course and in 1963 was given *Truncheon* as his first command. 1964 and 1965 were spent working in the Ministry of Defence before joining the nuclear ballistic missile submarine *Renown* in June 1966. He was subsequently promoted to Commander in June 1967 and served for a while as Flotilla Operations Officer on the staff of Flag Officer Submarines from November 1967 until the end of 1969. The following year saw a posting to the Royal Naval College at Greenwich before finally taking command of *Dreadnought* on 7 June 1971.

Dreadnought left Faslane on October 25 and sailed for Southampton where the submarine was scheduled to pay a courtesy visit to the thriving and bustling port. During her stay the ship's company was entertained by HMS *Wessex* the local RNR unit. She left Southampton on 1 November and headed back north to Faslane for more exercises in the Clyde including Exercise JMC 163. This exercise saw *Dreadnought* operating with *Onyx* and *Narwhal* and the guided missile destroyer *Antrim* and three frigates. More exercises followed and having successfully proved her worth, *Dreadnought* returned to Faslane for much needed repairs and maintenance on 10 November. November also saw a return visit to the submarine by HRH Princess Anne. On the 25 November the submarine had returned home to the Clyde and to Faslane Naval Base. The Princess had accepted an invitation to tour two of the base's resident submarines, *Dreadnought* and the Polaris nuclear ballistic missile submarine *Renown*. During her tour of *Dreadnought* Princess Anne's standard was flown from the submarines' mast as she was piped onboard accompanied by Commander Graham Rogers, CO of *Dreadnought*. Princess Anne was delighted to receive a gift from the ship's company of a paperweight fashioned into the shape of the submarine's badge.

Dreadnought was alongside throughout the Christmas period and into the new year when the nuclear submarine was slated to participate in a large naval exercise called Exercise Vendetta which also saw the use of *Osiris, Aeneas, Opossum* and *Ocelot* between 7 and 18 January 1972.

More exercises occurred throughout January and into February including Exercise Linkex with the frigates *Dido, Jupiter* and the guided missile destroyer *London* between 1 and 3 February. If the crew of *Dreadnought* had felt that they only went to cold climes they were in for a treat when it was announced that the submarine's next port of call would be Bermuda in the Caribbean. The island was duly reached on 4 February and ten days were spent in and around the island with many of the ship's company enjoying some well-earned rest on shore leave. All too soon for many it was back to work with Exercise Lantreadex with the aircraft carrier *Ark Royal* through until February 28. The following day *Dreadnought* slipped into the British Virgin Islands for a couple of days. The submarine left these beautiful islands on 2 March and headed north to visit the US Navy at Mayport in Florida between 3 and 8th.

On St Patrick's Day 1972 *Dreadnought* arrived back at Faslane for a maintenance period which was completed on 9 July. 2 to the 10 August *Dreadnought* was part of CANMARCOM operations in the North Atlantic and operated with Canadian forces visiting the port of Halifax in Nova Scotia for five days from the 10 August. The submarine stayed on the Eastern Atlantic seaboard for the next six days before returning home to Faslane on 24th.

Two days in the middle of September 1972 had been set aside for Exercise Highland Fling and *Dreadnought* was to play a major role in its successful outcome. The two days chosen were the 11 and 12 and *Dreadnought* operated with mine countermeasures vessels *Bildeston, Nurton* and *Iveston* and a frigate. Hardly had this exercise been completed than another commenced. Exercise Strong Express was a much larger event that also saw the integration of Canadian forces. British forces were led by the aircraft carrier *Ark Royal*, the assault ships *Fearless* and *Albion* and the helicopter cruiser *Blake*. These were backed up with two destroyers, six frigates, six mine countermeasures vessels and five support ships and two Canadian surface vessels. Underwater in addition to the *Dreadnought* the force comprised the nuclear powered *Conqueror* and the conventionally powered *Oberon, Alliance, Grampus, Rorqual, Walrus* and HMCS *Onondaga*. At the completion of the exercises *Dreadnought* headed back to Faslane where she arrived on 30 September. A few days later on 4 October *Dreadnought* headed south to Chatham Dockyard for an extensive maintenance period.

Chatham Dockyard would become a place that *Dreadnought* would come to know very well over the next decade. She arrived at Chatham Dockyard for the first time on 6 October 1972 at midday exactly.

On 27 November 1972 Commander John Patrick (Paddy) Bruce O'Riordan took command. He had joined the Royal Navy as a National Serviceman and as a Midshipman RNVR served in *Artemis*. On completion of National Service he asked if he could transfer to the RN. 'Their Lordships', in their wisdom, wrote saying that in view of his former Officer service he could enter the Royal Navy as an Able Seaman rather than as an Ordinary Seaman. Once commissioned as an Officer again he re-joined Submarines and served in *Sea Devil* (Mediterranean), *Alliance* (Home, South Africa and Far East including Confrontation) before becoming First Lieutenant of *Finwhale* (Home). After COQC he became the first lieutenant to take command of a Porpoise class submarine, *Porpoise* herself (Home and Mediterranean). After this and the Nuclear Course he went as the Executive Officer of *Courageous* building at Barrow and whilst there served for three months in a US nuclear submarine USS *Sunfish*.

Paddy O'Riordan recalls that "the *Dreadnought* team was a fun one, David Barraclough (same Training Class as Paddy O'Riordan) was XO, Tony Mayley (Engineer), Chris Field (WEO), plus Mike Samborne the son of *Dreadnought*'s first Captain was on board. The team also included Mike Chase, Gordon Leverett and 'trendy' Peter McKinley."

Dreadnought remained at Chatham Dockyard for the next four months until on 16 February 1973 she ventured out of the dockyard and into the waters of the River Medway once more. She returned home to the Clyde area on 19 February. Paddy O'Riordan remembers that "we completed a mini work up under Captain SM Third Submarine Squadron (SM3) and his able team of Keith Pitt, Bob Worlidge, Geoff Briggs, Andy McVean and the doctor, John Harrison, before heading for Gib and some spring sunshine." Gibraltar was reached on 12 March. Five days later the submarine entered the historic harbour at the base of the Rock and docked. She remained in the area until 9 April when she returned to the United Kingdom in time to celebrate the boat's tenth anniversary. The occasion saw the return of Peter Samborne to the submarine. *Dreadnought* completed a period of alongside maintenance on 3 June.

The Government policy of the early 1970's was to send 'balanced forces' on foreign deployments and accordingly *Dreadnought* was assigned to a group that comprised the helicopter cruiser *Tiger*, in which the Flag Officer Second Flotilla Rear Admiral Dick Clayton was flying his flag, and the frigate *Rhyl* on a Far Eastern mission in the summer of 1973. *Dreadnought* left Faslane on 11 June and joined the other members of the group en route. The *Dreadnought* was also joined by *Dido, Hermione* and two Dutch frigates *Limburg* and *Van Galen*. The force had been intending to travel to the Far East via the Suez Canal but with this closed the trip would involve the longer voyage around the Cape of Good Hope. *Dreadnought* and the other members of the task group reached Gibraltar on 8 June 1973. The submarine's crew decided to challenge the Army to a game of boule and soundly thrashed the pongos. Subsequently the ships operated together in the Atlantic on various exercises before crossing the equator.

The group reached South Africa on 3 July and when *Dreadnought* entered Simonstown Naval Base she became the first nuclear powered vessel to do so. As she approached the harbour she was escorted into port by the tug *De Neys* and the ex Admiralty launch *DL5* which had been re-christened from HL(D) *3939*. Rear Admiral Paddy O'Riordan recollects that her visit caused an enormous amount of media interest. "Their Defence (later Prime) Minister Botha visited the submarine and never removed his homburg – not even during lunch!" The crew's winning ways with boule continued when they beat a team from the crews of the South African French built submarines. *Dreadnought* and the other members of the force spent a relaxing if hectic few days in South Africa before leaving on 9 July. Further exercises had been organised with the South African and French Navies during the passage from South Africa to Singapore and these were actively participated in.

During this mission *Dreadnought* was the first vessel on the scene of a collision at sea and made a dramatic rescue of crew and passengers from the two ships involved. What they thought of being rescued by a Royal Navy submarine is sadly not recorded.

The warships were in the Malacca Straits on 26 July 1973 and received an SOS call when the two ships collided. *Dreadnought* found the 20,000 ton Cypriot registered tanker *Anson* embedded deeply into the hull of the Panamanian *Carnation* whose 35 man crew had taken to the vessel's lifeboats. The nuclear submarine surfaced and quickly rescued these men. Survivors from the *Carnation* stepped from the lifeboats across *Dreadnought*'s forward hydroplanes and to safety within the submarine. Rear Admiral Paddy O'Riordan's memories of this time were that the submarine leapt into action. "Hot food was provided for the survivors – first man down the hatch was Chinese and this was relayed to Chef and Tanky. It did not take long to rustle up 'chinky nosh' for 22 Cypriots who experienced their first Chinese meal." Meanwhile *Tiger* joined the rescue and launched three of her Sea King helicopters. Those rescued by *Dreadnought* were later transferred to the frigate *Rhyl* which was thought to be a much more suitable vessel for civilians.

27 July saw *Dreadnought* arrive at Singapore and the nuclear submarine was greeted with an amazing welcome of Chinese firecrackers when she berthed alongside the submarine depot ship *Forth* in the Johore Straits. Singapore has always been a favourite destination with Royal Navy crews and this time it was to be no exception with the crew enjoying many shore leave activities throughout their stay, which ended on the 20 August, three days later than planned after a severe storm forced the cancellation of the planned departure date of the 17 August. Back at sea Exercise Green Light with *Tiger, Rhyl, Hermione,* RFA *Tidespring* and RFA *Regent* took place over seven days ending on the last day of August.

"Frank Grenier," Rear Admiral O'Riordan continues, "earmarked to command the *Valiant,* joined us as part of his pre joining routine (CO's train CO's) to see how nuclear submarines should be run." Frank Grenier must have learned a lot onboard *Dreadnought* as he eventually became Flag Officer Submarines.

Later the Royal Navy task group entered Hong Kong. *Dreadnought* was accompanied into HMS *Tamar* by the helicopter cruiser *Tiger,* the frigate *Rhyl* and the RFA's *Regent, Tidespring* and *Tidereach.* It was a packed house in Hong Kong for the group's arrival. The Royal Navy force at the small naval base comprised five 'Ton' class patrol vessels which had recently been joined by the Australian frigate HMAS *Yarra* and also six Malaysian vessels.

The next scheduled port of call was to have been Sydney in Australia to participate in the official opening of the new Sydney Opera House. However, the Australian Government of the day had strong views on nuclear weapons and nuclear powered vessels and accordingly refused *Dreadnought* entry into Australian waters. The political fallout from this was that the entire British element of the task group decided not to go to Sydney and instead paid a visit to Guam between 11 and 25 September. At the time Rear Admiral O'Riordan noted "an eerie feeling as one approached Guam was to realise that the sea bed was six miles beneath." Guam being a US Navy Base had specialist facilities that could be used to repair any defects found in *Dreadnought*'s reactor and structure. None, however, were found and the American's were happy to give the reactor the 'once over'. 31 August to 5 September the group were welcomed by American forces in the Philippines at the US Navy Base in Subic Bay.

It was back to Singapore between 5 and 22 October before the submarine started her long voyage home to British waters. "We were back at Singapore for a maintenance period and once again we were joined by our wives. Besides maintenance for the submarine there was plenty of sport for the lads and a course at the Jungle Warfare School for 22 of the Ship's Company." *Dreadnought*'s return journey to the United Kingdom was via the paradise island of Mauritius between 6 and 9 November. "En route to the island, *Dreadnought* surfaced for the helicopter transfer of the well known broadcaster and RNVR Officer Richard Baker. His two-day visit included recording 'Start the Week' from my cabin at 400 feet and 30 knots. Our visit to Mauritius also coincided with Soviet celebrations to mark the October Rising. Rear Admiral Dick Clayton obtained approval from the Ministry of Defence for us to attend the party. As we bid farewell it was good to see the Soviet Naval Attaché festooned with Orders of Lenin, having amongst all his medals a lapel badge, "HMS DREADNOUGHT SUBMEN ARE SUPER". Ten days were also spent at Simonstown in South Africa with the submarine and group leaving the African sunshine on 26 November. *Dreadnought* and *Tiger* operated together in the surrounding waters and conducted Exercise Capex over four days until 30 November. At its completion the warships returned to

Simonstown until 3 December. The group then headed northwards and *Dreadnought* flying Admiral Clayton's flag arrived back at Faslane on 21 December 1973 and secured a place for herself in the history books. *Dreadnought*'s 194 day deployment was the longest patrol carried out by a Royal Navy fleet submarine. The MOD press men of the time were keen to promote some salient facts about *Dreadnought,* these were that the submarine had covered some 38,273 nautical miles and had only needed 40 gallons of diesel fuel for her generators during the whole six and a half month patrol. At the Press Conference once alongside the Admiral said, "Speaking as a non submariner it's the best thing I've ever done. I think it has great hopes for the future."

18 February 1974 saw *Dreadnought* at sea in company with two other Royal Navy nuclear submarines, *Valiant* and *Courageous* for Exercise Stem Turn. A few weeks later *Dreadnought* was undertaking trials off Campbeltown between 6 and 12 March. On 22 March *Dreadnought* pulled into Faslane and spent the remainder of the month exercising in the Clyde area. The final port visit of the commission was scheduled to be a return three day visit to the huge port of Southampton in Hampshire from the 4 to the 7 April. The next day *Dreadnought* arrived at Chatham Dockyard on 8 April 1974 at 13.45 for the process of refitting and refuelling to begin.

Chapter Four

Third Commission, 1974 -1980

The Royal Navy's early nuclear powered submarines required regular refuelling and refits and these meant that the enriched uranium cores had to be removed and replaced at about five year intervals. *Dreadnought*'s cores were replaced for a second time when the submarine was refitted at Chatham in 1974. The submarine would stay in the dockyard for a period of two and half years finally completing her refit in 1976. It is interesting to note that *Dreadnought*'s refit caused some considerable logistical and personnel problems at its commencement due to insufficient members of the Dockyard's workforce volunteering to work on the submarine. They feared for their health when working within the nuclear submarine's reactor compartments. The problem was eventually solved and *Dreadnought*'s refit continued without any further significant incidents. In an effort to ease some concerns Chatham Dockyards' Port Admiral, Rear Admiral Stephen Berthon met with the men working on *Dreadnought* in July 1974. The Admiral donned protective overalls and safety gear and then ventured into the reactor compartment of the hunter-killer submarine.

In October 1975, Chatham Dockyard's nuclear submarine refit facility was especially busy when *Conqueror* joined two other nuclear boats at the yard, *Dreadnought* and *Churchill* in dock. The process of refurbishing *Dreadnought* used many innovative techniques and equipment. This resulted in a major decrease in the amount of radioactivity within the hull. Indeed at completion of the refit some 40,000 gallons of radioactive material had been removed from *Dreadnought*. The refit was completed to schedule and slightly under budget with 50 days shaved off the refit by the construction of 'a shed' over the nuclear submarine during the refit to keep the weather out and to contain any avoidable release of material into the atmosphere.

Dreadnought was rededicated at Chatham on 9 October 1976 in a special ceremony which coincided with the 75th anniversary of the Royal Navy submarine service. At the rededication ceremony the guest of honour was Lord Fisher of Kilverstone, the grandson of Admiral of the Fleet Jacky Fisher. The ceremony was conducted by the Reverend M H G Henley, RN, Chaplain of Chatham Navy Base. He was assisted by the Reverend Father J Hine, Roman Catholic and the Reverend R H Norgate from the Church of Scotland and Free Churches.

Dreadnought received onboard her new commanding officer who was appointed to the post on 27 October 1975. Commander Hugh Kenneth Percival Michell had overseen the completion of the refit and the start of an extensive series of dock trials and subsequent sea trials to prove the success of the refit and to re-acquaint crew members with the design and operation of the Royal Navy's first nuclear powered submarine. Commander Michell started his Royal Navy career in the mine warfare vessel *Puncheston* before serving onboard submarines including *Explorer, Osiris, Finwhale* and *Rorqua*l.

Dreadnought nudged her way through the North Lock of Chatham Dockyard's Bullnose at 09.00 on the morning of 22 October 1976 and commenced post refit trials.

1977 was HRH Queen Elizabeth II's Silver Jubilee year and to celebrate the Royal Navy organised a massive review in the Solent to mark the occasion. As one of the Royal Navy's most modern and potent weapon systems *Dreadnought* was given pride of place in the line of ships. The Queen who launched the submarine seventeen years earlier in 1960 was said to be pleased to see the submarine in the fleet as she sailed past in the Royal Yacht *Britannia*. For three days between 24 and 27 June 1977 *Dreadnought* was positioned in the middle of an eight ship line of submarines between *Oracle* and the nuclear powered hunter-killer *Churchill*.

Dreadnought was, once again, in autumn of 1977 detailed to go east of Suez, but there was to be hitch to the plans. In October 1977 *Dreadnought* was refused permission by the Egyptian government to transit the Suez Canal and instead had to make the long journey round the Cape of Good Hope. *Dreadnought* was part of a large force including the cruiser *Tiger* that had planned to sail to the Persian Gulf with amongst other aims that of selling arms to Iran. The submarine had been due to visit the new naval base at Bandar Abbas on a 'goodwill visit'.

Later that year during a period of economic and political strife in Argentina the Labour government of Jim Callaghan heard a rumour from British Intelligence of a possible planned invasion by Argentine forces of the Falkland Islands. Argentina has for many years claimed sovereignty of the islands. The British government decided that a powerful task force should be assembled and forward deployed to the South Atlantic to forestall any such plans. In November 1977 *Dreadnought* was despatched as part of Operation Journeyman along with the frigates *Alacrity*, *Leander* and *Phoebe* and the supply ship RFA *Resource* and tanker RFA *Olwen*.

The British and Argentine forces never came to blows but the tension was palpable amongst both sets of crews. At one point *Dreadnought* is alleged to have surfaced near to the RFA *Pearleaf* causing some degree of consternation. *Dreadnought* had been under orders not to surface during the mission. It was also suggested at the time that British Intelligence operatives working in Argentina suggested to the Argentines that *Dreadnought* was operating off the Argentine coastline near to her major naval ports waiting and watching ready to attack if the order was received from the Ministry of Defence in London. In the end, the tension died away when the military junta in Buenos Aires turned their attention to domestic matters. The question of possession of the Falklands would, however, return to haunt politicians in London just five years later during the 1982 conflict. *Dreadnought*, meanwhile, returned from the South Atlantic and resumed her normal pattern of operations. She did, however, bring back with her valuable information about the oceanographic conditions of the South Atlantic that would prove useful during the Falklands conflict. *Dreadnought* upon her return to UK waters spent many months on trials and crew rest periods usually alongside at Faslane.

In December 1977 *Dreadnought* was back in the Mediterranean and paid a visit to Thessaloniki in Greece. During this visit the nuclear submarine heard that the Sadler's Wells Ballet Company were in town on a tour of Greece. Invitations were quickly organised for an equally quickly organised cocktail party onboard and the company's Prima Ballerina Margaret Barbierie accepted the invitation and toured the submarine. Less exotic ports of call for the submarine also included a visit late in 1978 to the port of Hull.

At 09.45 on 26 January 1979 *Dreadnought* once more entered the safe confines of Chatham Dockyard. She had returned to the nuclear submarine refit complex at Chatham Dockyard for what was called a five month mini refit. Indeed *Dreadnought*'s DED (Docking and Essential Defects) was the first to use the new nuclear submarine refitting facilities provided in No 5 dock. Many of the workers at Chatham Dockyard were surprised when the extent of the work on *Dreadnought* was published. For a mini refit lasting only five months there was an unusually high amount of restorative and improvement work to be carried out on the submarine. Amongst many items on the agenda were a complete change of *Dreadnought*'s glass reinforced plastic dome for the Type 2001 sonar. This particular repair was a first of its kind for Chatham Dockyard. Additionally *Dreadnought* received repairs to her casing fairings and an extensive upgrade was carried out on the hunter-killer's reactor instrumentation to bring it up to the new nuclear safety standards.

The only real problems encountered during this refit were due to the American nature of *Dreadnoughts*' reactor compartment and its associated equipment. Many of *Dreadnought*'s American components were already being withdrawn from service on American boats of similar age and with fewer and fewer parts available for earlier nuclear reactors a spare parts supply problem started to become apparent and at times caused the refit schedule to slip slightly.

Commander Johnny P Clarke MBE joined the submarine on 22 May 1979. A little under a month later on 11 June 1979 *Dreadnought* had completed this 'mini refit' and again entered the River Medway to begin another patrol.

On 14 February 1980 *Dreadnought* was the scene for another small piece of history being made. The submarine was off the Isle of Arran and carried with her a team of BBC reporters, cameramen and technicians who had joined her for a live television broadcast. This was the first time that such a broadcast had been conducted from a nuclear powered submarine anywhere in the world. The BBC team from the 'Pebble Mill at One' programme had set up a transmitter on the nearby Isle of Arran and the live show was transmitted via this land site. The programme included interviews with the Flag Officer Submarines, Rear Admiral 'Tubby' Squires and the commanding officer of *Dreadnought* Commander Johny Clarke and also members of the crew. *Dreadnought* had also welcomed onboard the singer Lyn Paul who recorded some songs prior to the live transmission.

Another refit occurred in 1980 and was again carried out at Chatham Dockyard. It was, however, the Defence Review of 1981 that spelt the end of *Dreadnought's* operational career. She was older and noisier than the newer Valiant and Swiftsure class boats in the fleet and with only a limited hull life remaining, the decision was taken to decommission *Dreadnought*. In retrospect the decision was a correct one as the vessel was becoming obsolete in relation to newer British, American, French and Russian nuclear submarines.

Dreadnought was at Faslane in August 1980 when her reactor had to be shut down when a major fault was discovered in a backup cooling system during routine trials. At the time no one believed that it would be the last time that *Dreadnought* would sail under her own power. A Navy spokesman at the time said of the incident, "It's the sort of situation we would rather like to avoid" She stayed at Faslane for some weeks until the decision was taken to tow her to Devonport where the fault could be corrected in dry dock. The reason for not sailing under her own power was to "avoid putting unnecessary strain on the other backup cooling system" The tugs that had taken the aircraft carrier *Ark Royal* to Cairn Ryan for scrapping were used to tow the stricken submarine on 30 September 1980. For the journey south *Dreadnought* was to be escorted by the frigate *Minerva* and together with the tugs the submarine and escorts arrived at Devonport on 3 October 1980. The original plan had been to refit the submarine but the decision was later taken to decommission her instead as the cost to keep her in commission was considered too high. In March 1982 Mr Peter Blaker, Minister of State for the Armed Forces answered a Parliamentary question about the future of *Dreadnought* he said, "This submarine is noisy by modern standards, and even after a refit would only have a limited operational capability in her primary role. The cost of this refit would have been in excess of £70 million.

Dreadnought was then prepared to go to the Standby Squadron at Chatham Dockyard which was also selected to complete Britain's first complete nuclear de-fuelling. Quite an achievement for a dockyard that was due to be closed by the 1981 Defence Review. Originally a refit on the seven year old *Sovereign* was allocated to the yard but this was cancelled and replaced by the multi million pound work on *Dreadnought*.

Dreadnought left Plymouth under tow of the ocean going tug RMAS *Rollicker*. The submarine was essentially inert, her reactor was shut down and onboard there was just a skeleton crew. As befits such a voyage, *Dreadnought* and *Rollicker* were given a frigate escort all along the English Channel by the Type 21 frigate, *Active*. *Active* left the two vessels as they neared the Thames Estuary. The tug towed *Dreadnought* past Sheerness and into the Medway Estuary with its twists and turns before approaching Chatham Dockyards' Bullnose on 9 October 1981 where RMAS *Rollicker* handed over the tow to the dockyard tug RMAS *Felicity*.

RMAS *Felicity* pushed and pulled the nuclear submarine through Basins 3 and 2 and finally into No 1 Basin. *Dreadnought* was then edged into No 5 dock, which remained flooded while No 7 dock in the Submarine Nuclear Refit Complex was prepared for her dry-docking. *Dreadnought*'s refit was to be one of the most complex tasks ever asked of Chatham Dockyard - the complete defuel and stripping of a nuclear submarine. *Dreadnought* entered Chatham's No 7 dock with an expected start date the work of 1st March 1982. The decommissioning of *Dreadnought* would involve up to 150 men at Chatham Dockyard and cost £9 million and take nine months. In the event the refit cost considerably less at £4.3 million. The refit comprised five main areas. A complete de-fuel of the submarine, sealing of the Reactor Plant. Special attention was paid to preserving the hull for its expected long period of inactivity in reserve, these included coating the hull to try to prevent boring marine creatures from attaching to the steel of the hull and to prevent corrosion generally. De-equipping was done to remove any useful materials and equipment suitable for reuse on other Royal Navy submarines. Finally arrangements were made for towing and mooring the nuclear submarine when the time arrived to take her to Rosyth. In the April 1983 edition of Chatham Dockyard newspaper 'Periscope' General Manager of Chatham Dockyard, Mr Alan Kettle said, "It has been a unique work project and completing it was a significant milestone for Chatham. A lot of people have done a lot of hard work."

What to do with the hulk of the still radioactive *Dreadnought* was discussed at high level in the Government. There was pressure to move the submarine to one of the Royal Navy's surviving naval bases at, Devonport, Portland or Portsmouth. The favoured option was, however, Rosyth where there were excess facilities and berths for the submarine. This possibility caused the resident Member of Parliament for Dunfermline to ask the government of its intentions. He was very concerned about the possibility that Rosyth would become "a scrapyard for nuclear boats." Another option that appeared at the time was the possibility of preserving the *Dreadnought* as a memorial and museum ship. At least two requests to the Ministry of Defence were made regarding this option, one from Merseyside and the other from Morecambe. A spokesman from the MOD at the time, said, "Some interest has been shown in preserving her for public display and the MOD are prepared to consider any proposals sympathetically." Unfortunately, what plans and enthusiasm there were for *Dreadnought* to go on display, dissipated.

Dreadnought was un-docked from the nuclear refit complex at Chatham and floated out in No 1 Basin where she stayed in the Reserve Fleet in Category C5 (out of service 'due to material and/or training deficiency') with an availability of over 30 days. *Dreadnought* was formally decommissioned from the Royal Navy on 14th March 1983. A little under a month later the redundant submarine was towed from Chatham on 13th April 1983 by the RMAS *Roysterer* and headed to Rosyth.

Rosyth Dockyard in Scotland had been selected as the preferred location to 'store' *Dreadnought* before a decision was made on her disposal. But this controversial submarine continued to make headlines but for all the wrong reasons. The local members of CND in Scotland began protesting about *Dreadnought*'s arrival. One was quoted in the local newspapers as saying, "We are angry about this bloody thing being dumped here. It is by nature of an experiment – they are going to leave it there until it rots, and there are no ill-effects"

Dreadnought arrived in Scotland on Saturday 16 April 1983, under the control of two RMAS tugs. As she was towed up the River Forth and under the mighty Forth bridges she sailed straight into an industrial dispute with disgruntled workers at Rosyth Dockyard. The workers were protesting about the possible use of Rosyth as "A nuclear dumping ground'. 1500 workers at the yard of the Shipwrights Section of the General and Municipal and Boilermakers Union voted to 'black' the submarine. Mr Bert Lumsden, secretary and convener of the Rosyth Dockyard branch of the EETPU said to the Dunfermline Press that "The blacking is our way of protesting at *Dreadnought* being dumped at Rosyth. There is no CND involvement or anything like that."

Upon her arrival at Rosyth *Dreadnought* was, according to Mr David Neilson, Civil Secretary to the Port Admiral at Rosyth, "sealed to prevent unauthorised entry and *Dreadnought* will be berthed in an area used for other ships awaiting disposal."

At first it was expected that the obsolete radioactive hull would be sunk at sea in as deep a location as could be found in the ocean. It was proposed that the submarine should be filled with concrete and sunk 400 miles off the coast of Cornwall. This policy was severely criticised by organisations such as Greenpeace and Friends of the Earth who thought such a plan would be an ecological disaster waiting to happen. The Ministry of Defence begrudgingly agreed and *Dreadnought* has remained at Rosyth since her arrival there in April 1983. She is periodically taken into dry dock to check for corrosion and the state of her hull. These inspections cost the taxpayer several million pounds on each occasion and now in 2002, *Dreadnought* has been joined by other redundant Royal Navy nuclear submarines in the same condition.

Technical Details

Displacement:	3000 tons standard; 3500/4000 tons submerged
Dimensions:	81metres x 9.8m x 7.9m
Machinery:	1 shaft nuclear propulsion based around a single S5W reactor plus geared steam turbine. This produced 15,000shp. There was also diesel electric auxiliary propulsion available for emergencies.
Speed:	15 knots surfaced and 28 knots submerged
Armament:	6x21 inch torpedo tubes (24 torpedoes are available)
Sensors:	Search and navigation radars; sonars 2001, 2007 and 2019.
Complement:	88
Pennant Number:	S101
Laid down:	12 June 1959
Launched:	21 October 1960
Completed:	17 April 1963

Commanding Officers

Name	Date joined *Dreadnought*
Commander B. F. P. Samborne	28 July 1960
Commander J. D. E. Fieldhouse	27 July 1964
Commander Peter Cobb	6 April 1965
Commander Alan Kennedy	6 January 1969
Commander Graham Rogers	7 June 1971
Commander J. P. B. O'Riordan	27 November 1972
Commander H. K. P. Michell	27 October 1975
Commander J. P. Clarke, MBE	22 May 1979

Previous HMS Dreadnoughts

The first Royal Navy vessel to have the name *Dreadnought* was a 40-gun warship built in 1553. The second *Dreadnought* was present during the Battle of the Armada. She was built in 1573 and another highlight of her career was the battle of Cadiz in 1596. This ship served continuously until broken up in 1645. The next *Dreadnought* was previously named *Torrington* and re-christened as *Dreadnought* in 1660. She had 64 guns, which came into their own on three separate occasions; during 1666 at the battle of Lowestoft, The Four Days' Battle and for a third time at the battle of Orford Ness. In 1672 *Dreadnought* fought in the battle of Sole Bay and in the following year she was again in action at Schooneveld and again at Texel. After another twenty four years service the ship foundered in 1690.

1691 saw the next *Dreadnought* enter Royal Navy service as a 4th rate. This *Dreadnought* saw action on two occasions. The first was the battle of Barfleur in 1692 and the second in 1718 at Cape Passaro. This vessel survived until broken up in 1748.

In the same year a prize sloop called *Dreadnought* was captured and later sold.

However, another *Dreadnought* had entered service in 1742 as another 4th rate warship. 1757 saw the name *Dreadnought* secure another battle honour at the battle of Cape Francois in 1757. For the next twenty seven years she sailed the world until sold for breaking up in 1784.

The next vessel to bear this historic name was called *Dreadnought's* prize but was sold out of commission in 1798.

Three years later in 1801 a 2nd rate warship was commissioned bearing the name of *Dreadnought*. This *Dreadnought* was part of the fleet assembled by Admiral of the Fleet, Lord Horatio Nelson for the attack against the French and Spanish Fleets at the Battle of Trafalgar in 1805. In 1827 she became a hospital ship and continued in service until broken up in 1857.

That year saw the previously named 1st rate *Caledonia* renamed as *Dreadnought* and for the next eighteen years she sailed the world as a hospital ship.

A steel battleship was constructed in 1875 to replace the previous vessel of the name. She continued to exist until 1908.

Two years previously had seen the introduction into the Royal Navy of one of the most famous battleships in history. *Dreadnought* was of a completely revolutionary design that made all previous battleships obsolete almost overnight. She was armed with 12″ guns that could train in any direction and the propulsion plant was also of a revolutionary design that used turbines for the first time. This gave *Dreadnought* a speed of 21.5 knots on trials. This itself was remarkable in that *Dreadnought* was already some four or five knots faster than previous battleships. Also interesting is that the construction of this novel and revolutionary warship took just one and a half years from her keel laying within Portsmouth Dockyard to her completion. In many ways *Dreadnought* was a catalyst towards the arms race that led to two world wars in the 20th century. *Dreadnought* was to become the flagship of the 4th battle squadron and whilst she missed the Battle of Jutland she did sink the German U-boat *U29* by ramming her in the North Sea. *Dreadnought* was in August 1917 considered to be in relatively poor state and was put into the Reserve Fleet. Sold in 1921, she was broken up in 1923.

The name *Dreadnought* lapsed for a period of thirty nine years until Britain's first nuclear powered submarine was launched by Her Majesty Queen Elizabeth.

Bibliography

Books

Conway's All the World's Fighting Ships 1947-1995, 2nd Revised edition, Conway Maritime Press, 1995.

Janes Fighting Ships, Janes 1958-1982

HMS Dreadnought, the First Five Years, Rolls Royce & Associated Ltd, 1968

Newspapers & Magazines

Periscope (The Chatham Dockyard internal newspaper 1960-1981)

Ships Monthly

Sea Breezes

Warship World

Archives

Royal Naval Museum, Portsmouth

Royal Navy Submarine Museum

Chatham Historic Dockyard

Plymouth City Archives

Portsmouth City Archives

Public Records Office, Kew

INDEX

Y